Visualizing Data Patterns
with Micromaps

CHAPMAN & HALL/CRC
Interdisciplinary Statistics Series

Series editors: N. Keiding, B.J.T. Morgan, C.K. Wikle, P. van der Heijden

Published titles

AN INVARIANT APPROACH TO STATISTICAL ANALYSIS OF SHAPES	S. Lele and J. Richtsmeier
ASTROSTATISTICS	G. Babu and E. Feigelson
BAYESIAN ANALYSIS FOR POPULATION ECOLOGY	Ruth King, Byron J.T. Morgan, Olivier Gimenez, and Stephen P. Brooks
BAYESIAN DISEASE MAPPING: HIERARCHICAL MODELING IN SPATIAL EPIDEMIOLOGY	Andrew B. Lawson
BIOEQUIVALENCE AND STATISTICS IN CLINICAL PHARMACOLOGY	S. Patterson and B. Jones
CLINICAL TRIALS IN ONCOLOGY, SECOND EDITION	J. Crowley, S. Green, and J. Benedetti
CLUSTER RANDOMISED TRIALS	R.J. Hayes and L.H. Moulton
CORRESPONDENCE ANALYSIS IN PRACTICE, SECOND EDITION	M. Greenacre
DESIGN AND ANALYSIS OF QUALITY OF LIFE STUDIES IN CLINICAL TRIALS, SECOND EDITION	D.L. Fairclough
DYNAMICAL SEARCH	L. Pronzato, H. Wynn, and A. Zhigljavsky
GENERALIZED LATENT VARIABLE MODELING: MULTILEVEL, LONGITUDINAL, AND STRUCTURAL EQUATION MODELS	A. Skrondal and S. Rabe-Hesketh
GRAPHICAL ANALYSIS OF MULTI-RESPONSE DATA	K. Basford and J. Tukey
INTRODUCTION TO COMPUTATIONAL BIOLOGY: MAPS, SEQUENCES, AND GENOMES	M. Waterman

Published titles

Chapman & Hall/CRC
Interdisciplinary Statistics Series

Visualizing Data Patterns with Micromaps

Daniel B. Carr

George Mason University
Fairfax, Virginia, USA

Linda Williams Pickle

StatNet Consulting, LLC.
Gaithersburg, Maryland, USA

CRC Press
Taylor & Francis Group
Boca Raton London New York

CRC Press is an imprint of the
Taylor & Francis Group, an **informa** business

A CHAPMAN & HALL BOOK

Chapman & Hall/CRC
Taylor & Francis Group
6000 Broken Sound Parkway NW, Suite 300
Boca Raton, FL 33487-2742

© 2010 by Taylor and Francis Group, LLC
Chapman & Hall/CRC is an imprint of Taylor & Francis Group, an Informa business

Library of Congress Cataloging-in-Publication Data

Carr, Daniel B.
    Visualizing data patterns with micromaps / Daniel B. Carr, Linda Williams Pickle.
        p. cm. -- (Chapman & Hall/CRC interdisciplinary statistics series)
    Includes bibliographical references and index.
    ISBN 978-1-4200-7573-1 (hardcover : alk. paper)
    1. Mathematical statistics--Graphic methods. I. Pickle, Linda Williams. II. Title. III. Series.

    QA276.3.C375 2010
    519.5--dc22   2009043229

**Visit the Taylor & Francis Web site at
http://www.taylorandfrancis.com**

**and the CRC Press Web site at
http://www.crcpress.com**

To Jean and Jim

And to our grandchildren

Taishan, Ethan, Serenity, Adeline, and James; Naomi and Steven

who have inspired us by taking their own next steps—
some by just learning to run and others by learning to graph data

Contents

Preface

Breaking out of the classic mold, when the breakout is effective, takes us much further than improvements within that model.

—John Tukey (Tukey 1988b)

After over fifteen years of research and trial and error, micromap designs have evolved to the point where they are slowly finding their way into mainstream statistical visualizations. Now seems to be a good time to pull all of the work together into a book in order to introduce micromaps to a wide range of people interested in visualizing their data. Our intent is not only to describe these graphics but also to present the research of others in cognitive psychology, statistical graphics, computer science, and cartography who laid the foundation for our methods. By understanding why we favor particular design elements in micromaps, you should be better able to tailor your micromap designs to be more effective. If we have missed opportunities or made compromises not to your liking, you can use the guidance here, do it your way, and do better. Thus, we have written this book for anyone who wishes to explore the statistical and geographic patterns in their data simultaneously and for designers of visualization tools that will support visual exploration and communication of patterns in maps.

We begin in Chapter 1 by illustrating the three main types of micromaps. The next chapters summarize the research behind the designs (Chapter 2) and explain how these research findings can be applied to micromap designs in general (Chapter 3). These chapters are most important for designers but should provide helpful background for choosing design options even by those who just want to look at their data. We go into detail about the specific micromap designs in the next three chapters on linked, conditioned, and comparative micromaps. Chapter 7 brings it all together by applying all three techniques to the same data set to compare and contrast their purposes, limitations, and strengths.

We hope this book will be a part of courses taught by many different disciplines. Dan has used micromaps as a part of his classes on statistical graphics and data exploration, and scientific and statistical visualization for many years. His classes emphasize redesign of poor graphics and tables, application of graphics to new data, presentations, and written papers. Students applying the linked micromap designs to data from their own work have often received positive responses from their employers and this has influenced a few careers. Students passionate about their data almost always do well in class. Some students who are well trained in statistical methodology but do not have a strong application area preference need to be guided to real-world data. In this case the federal agencies offer a wealth of possibilities. While it would be easy and fun to base a whole course on micromaps, we primarily see adding micromaps as a way to enrich many existing courses.

We are providing many resources on the book website, http://www.crcpress.com/product/isbn/9781420075731, and www.statnetconsulting.com/micromaps, which contains many boundary files and data sets. The software includes CCmaps (Java™ (Sun Microsystems)) for conditioned micromaps along with working examples for the United States by state, United States by county, and each individual state by county. One large folder of data examples addresses cancer statistics, risk factors, and demographic data. We have found that people often enjoy exploring their own data, so it is relatively easy to import other data sets into these sample programs. Some people do enjoy exploring data provided by others, so there are some contributed examples, such as species biodiversity on a hexagon grid for the Mid-Atlantic states from Denis White. Examples from other sources include the beginning of an alcoholism study for zip code regions in a city. We anticipate that the data set collection on the website will grow as we and others continue to contribute examples.

The website also contains R functions and scripts for linked micromaps and comparative micromaps. The data provided will allow re-creation of the corresponding new examples for this book. A Java-based interactive version of linked micromaps is available directly from the National Cancer Institute at http://gis.cancer.gov/tools/micromaps/. A dynamic, Java-based implementation of comparative micromaps, called TCmaps, will be available on our website. The website may contain other implementations if students choose to make them available.

Our collaboration arose from joint work over the past fifteen years. Dan's promotion of visualization methods resulted in new ways of linking regional statistics to maps in Linda's 1996 atlas of mortality and many later geospatial projects. The need of analysts at the National Center for Health Statistics (NCHS) and later the National Cancer Institute (NCI) to examine geographic patterns and their associations to multiple risk factors led to Dan's new conditioned micromap design and refinements of linked micromap designs. Dan spent a sabbatical year working with Linda in the Statistical Research and Applications Branch at the National Cancer Institute in 2001, where he further focused his thinking about communicating cancer statistics through visualizations. He was occasionally frustrated at the length of time that it took for a commercial-grade web application to be completed. However, he was impressed by how well the final product addressed the communication needs of a wide audience, a task more complex than developing software to demonstrate proof of concept. He was aware that some people misunderstood the horizontal color linking in linked micromaps at first encounter, but he did not fully appreciate how much the order of the panel columns mattered to beginning users. Extensive usability studies of the linked micromaps led by Sue Bell demonstrated, for example, how important it was to put row (state) labels on the left of the linked micromap panels. The NCI/CDC State Cancer Profiles website with linked micromaps became a wildly successful application for cancer statistics communication. Cancer epidemiologists, cancer control specialists, cancer registry staff, and cancer advocates all have utilized this site to understand the patterns of cancer in their local areas.

We are glad to see these designs being used in this way. However, the designs belong to everyone, not to any one discipline, age group, or nationality. Our tour of micromaps is conducted in English and examines data from a modest number of fields, such as health, environment, and demography. However, we are glad to see micromaps being produced by people in other nations, using different languages and addressing different topics. Designs should be revised in terms of reading directions and other cultural differences and as others contribute insights and improvements. Over time these newer and better designs can help people to understand, reason, and improve the quality of their life.

Micromaps follow in a long line of graphical innovations. To paraphrase John Tukey's quote above, sometimes you need to look at data in a whole new way in order to see something new and different. This is what we have tried to do with micromaps. There are many to thank for leading the way. These include pioneers in statistical graphics such as John Tukey, Bill Cleveland,

and Edward Tufte. Dan also thanks many others for their guidance and encouragement. From Pacific Northwest National Laboratory, these include Wes Nicholson for his vision of large data set analysis and computer scientists Jim Thomas and Rik Littlefield, who opened doors to the rich world of computer graphics and user interfaces. Special thanks go to Tony Olsen (EPA), who sought graphics to show results from environmental sampling and became the catalyst who sparked the idea of linked micromaps. In terms of human perception and cognition, Dan thanks Steven Kosslyn for his 1994 book on graph design. This provided Dan's first inkling that perceptual grouping had been studied and that groups of four had merit over Dan's intuitive groups of five. This led to Dan's further study of the cognitive literature and later to valuable insights from Colin Ware's book and discussions with Lee Wilkinson. From the world of cartography/geography Dan thanks Denis White, Alan MacEachren, and Cindy Brewer.

In terms of software, he thanks Andrew Carr for developing the first graphical user interface (GUI) for linked micromaps using Splus; Dan Rope for his related work with nViZn; Juergen Symanzek and David Wong for their effort toward the first web-based version; and Sue Bell (National Cancer Institute project leader), Jim Chen (George Mason University), Bill Killam (User-Centered Design), and James Cucinelli, David Eyerman, and Mike Balasanyan (Information Management Services, Inc.) for their roles in producing and usability testing the Java applet for the National Cancer Institute. Finally, he thanks former George Mason University students Duncan MacPherson, Yuguang Zhang, Yaru Li, and Chunling Zhang for their work on CCmaps and Chunling Zhang for her work on TCmaps. There are many others deserving of thanks.

Linda also thanks Douglas Herrmann, a cognitive psychologist at NCHS, and Alan MacEachren and Cynthia Brewer and others in the Pennsylvania State Department of Geography and at other universities for helping to conduct the series of cognitive experiments at the NCHS in the 1990s that looked into how map readers extracted statistical information. Thanks also go to the NCHS management (Manning Feinleib, Monroe Sirken, and Andy White), who had the vision for a new atlas and the desire to design it from a cognitive science perspective. Our long collaborative relationship with Alan MacEachren and his students plus the knowledge gained from the many other researchers involved in this NCHS atlas project broadened Linda's perspective beyond statistical graphics and led to greater consideration of the reader's visual and cognitive strengths and limitations in future map designs and communication efforts.

We also thank those who helped directly with this book. Our editor Bob Stern was enthusiastic about this book from the beginning, encouraging us when we were not sure whether we had enough material for an entire book. The early reviewers encouraged us to broaden the scope of the examples and made suggestions that greatly improved the readability of the book. Thanks also to Jim Pearson for stepping in to help with the R programming when deadlines were looming, for providing patient and skillful computer support to Linda 24/7, and for serving as a constructive critic of our explanations.

Finally, we thank our families for their patience and for keeping the households running while we buried ourselves in this project. Thanks especially to our respective grandchildren, who provided comic relief and much-needed mental breaks along the way.

The Authors

Daniel B. Carr, PhD, is professor of statistics at George Mason University. He received his PhD from the University of Wisconsin–Madison in 1976. After this he spent fourteen years as a research scientist at the Pacific Northwest National Laboratory gaining multi-disciplinary experience by addressing topics in biological and physical sciences, national security technology, and the analysis of large data sets. At Mason he has taught statistical graphics classes to hundreds of graduate students coming from many disciplines, and promoted the use of statistical graphics in federal agencies via collaborative research with visionary agency staff. He is a Fellow of the America Statistical Association and recognized for his contributions in statistical graphics.

Linda Williams Pickle, PhD, was a mathematical statistician at the National Center for Health Statistics and the National Cancer Institute before retiring in 2007. She has published extensively about spatial patterns of disease in the medical and statistical literature, including three national atlases of mortality. Now, as the principal of StatNet Consulting, LLC, and adjunct professor of geography and public health sciences at the Pennsylvania State University, she specializes in spatial statistical models and spatial data visualization methods. Dr. Pickle received her PhD in biostatistics from the Johns Hopkins University in 1977. She has received numerous awards from the NIH and CDC and is also a Fellow of the American Statistical Association.

1 An Introduction to Micromaps

Welcome to the world of micromaps, graphics that link statistical information to an organized set of small maps. We are eager to take you on a tour that explores the many design variations of micromaps and their continually expanding variety of applications.

1.1 INTRODUCTION

The primary purpose of micromaps is to highlight geographic patterns and associations among the variables in your data set, graphically integrating these two aspects of the data. Micromaps can be designed to communicate known patterns to others or to allow you or your audience to explore data. For exploring data, you can go beyond a simple description of patterns and use your knowledge of local conditions to generate hypotheses about plausible causes of the observed trends and relationships. Sometimes the micromaps uncover or communicate unexpected patterns that serve as powerful evidence to motivate action.

Micromaps are also a visual medium for telling stories. Just as journalists organize their information to discover and communicate their stories, analysts can organize their data to discover and communicate the stories in the data by using micromaps to answer the questions who, what, when, where, and sometimes why. For example:

- Who has the highest unemployment rates— men or women, Whites or African Americans, younger or older people?
- What is the range of unemployment rates?
- When did the trend in rates last change?
- Where are the rates highest?
- Why do these geographic patterns and trends exist? Can anything be done about them?

The data represented by micromaps are often statistical summaries. Even though the original data set may seem overwhelming in size and complexity, most people can obtain a reasonable understanding of the underlying data when these summaries appear in well-designed graphics. This is where micromaps excel— the basic visual encodings and implied comparisons are easy for most people to learn and the maps provide a context that helps bring the statistics to life. We can see how measures of health, environment, education, or agriculture vary from region to region.

The plots can include any of a number of statistical summaries to convey key features, such as point estimates, ranges, means with confidence intervals, or other distributional summaries. While some numerical literacy is important to understand these concepts, we want to stress that micromaps can help people to become more comfortable with statistics and are not limited to use by those trained in statistics. In fact, a nonstatistician who is familiar with the local data will often have the most insight into the displayed patterns. On the other hand, it would be wise to consult with someone knowledgeable about the data set early in the analysis, prior to drawing conclusions about the patterns revealed by the micromaps. Most people who study or present graphics are well intentioned, but may be unaware of quirks in the study design, data collection methods, variable coding, and file preparation that could lead to misinterpretations of the graphed data. Further, our eyes are so good at spotting patterns that we often perceive patterns in what is really random variation that we call noise. Just as flipping a fair coin can result in ten heads in a row, random variation in data can produce apparent clusters, associations, or other patterns. Good statistics can not only help us see patterns through the noise of random variation but also warn us against getting too excited about patterns that may well be due to chance.

We introduce the three major categories of micromaps in this chapter: linked, conditioned, and comparative. The micromap designs are very flexible, as we hope to illustrate by the many examples in this book. The fundamental layout for micromaps is a set of thoughtfully organized rectangular panels that typically include small maps. However, we begin with the direct precursors of linked micromaps, multiple panel plots called row-labeled plots that have no maps at all.

1.2 ROW-LABELED PLOTS

By 1994 the simple and effective dot plots of Cleveland and McGill (1984) had been available for a decade but

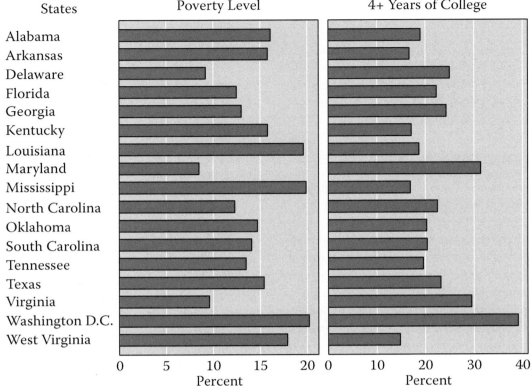

FIGURE 1.1   A row-labeled plot with horizontal bars of college education and poverty for southern U.S. states, 2000 Census, sorted by state name.

were still hard to find in federal publications. Dan coined the term *row-labeled plots* and developed both refinements and extensions to dot plots in his effort to promote statistical graphics and to broaden the audience for federal statistics (Carr 1994). He argued that the historical archival role served by tables had been largely replaced by electronic digital storage and for the majority of people federal communication efforts would be better served by graphics than tables. To illustrate, he redesigned tables commonly used to display federal statistics as visually simple but effective statistical graphics. The first linked micromap plots emerged as an extension of these row-labeled plots (Carr and Pierson 1996).

The underlying layout for a row-labeled plot is the same as a two-way data table, sometimes referred to as a flat file or spreadsheet layout. Each row corresponds to a different observational unit such as a state or county. (For simplicity, we will refer to any of these geographic units generically as a *region*, but keep in mind as you read this book that *region* can refer to any spatial representation.) Each column corresponds to a different variable. The

leftmost column is special in that it consists of labels for the observational units. The row-labeled design encodes the statistics in the remaining columns using familiar statistical graphics such as dot plots, bar plots, and box plots. Horizontal alignment links the information about each observed unit from the row label on the left to the last plotted statistic on the right—hence the name row-labeled plots. In row-labeled plots three features enhance the simple two-way layout—grid lines, sorting, and partitioning of rows—which we will demonstrate by example.

Figure 1.1 illustrates this design with Census 2000 data for poverty and college education for states (including the District of Columbia) in the South Census Region (U.S. Bureau of the Census 1994). This is a common plot design, with horizontal bars representing the values of the two variables and the rows sorted by state name. The first design feature to note is the use of grid lines in the background, which is good for comparing regions' values relative to a reference value or to other regions and for reading approximate values for a single region.

Poverty and Education in Southern U.S. States, 2000

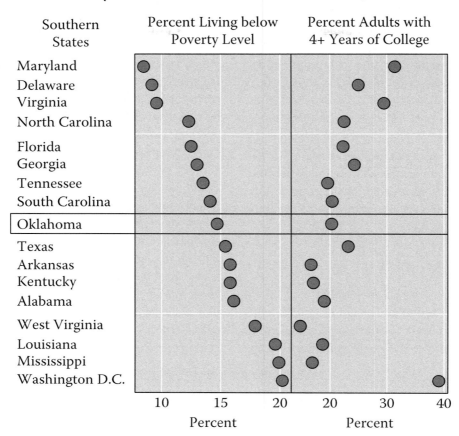

FIGURE 1.2   Figure 1.1 redrawn as a row-labeled dot plot. Sorting by poverty level provides a much clearer picture of the association between the poverty and education than the original alphabetic state sort.

For example, we can see that the District of Columbia (DC), Louisiana, and Mississippi each have about 20% of their residents living below the federal poverty level and that DC, Maryland, and Virginia have the largest percentages of college-educated adult residents.

Other patterns are more difficult to discern from this design. How might we improve it? We offer three improvements in the revised Figure 1.2. By changing from bars to dots, the reader can better focus on the variation in the states' values because the axis does not need to include 0. This provides better resolution by zooming in to the data range. The second design feature to note is that the rows are sorted. While alphabetic sorting based on the row labels is an option, rows are often sorted by one of the plot variables in order to bring similar values together, making pattern comparisons both across the columns and down the rows easier. Sometimes we use several variables to create an index as the basis for a rank ordering of rows (Carr and Olsen 1996; Cleveland 1985).

The third design feature to note is the partitioning of the sorted rows into small groups with similar ranks. This perceptual grouping by using horizontal lines in the background creates rows and columns of panels, each with modest content. The purpose is to facilitate local focus and to help the reader scan across the rows accurately. Comparison of columns within a row can reveal local associations among the plot variables, and comparison of rows within a column can show patterns in the distribution of values of a single variable. Figure 1.2 provides an example that puts states into groups of four. Since there is one extra state, the design calls attention to the state in the middle, i.e., the state with the median value.

What types of questions does this revised design help us answer? We might ask whether the data suggest an inverse association between poverty and education, as we might expect. The answer is yes. The revised row-labeled plot (Figure 1.2), now sorted from low to high poverty level, shows a strong inverse association, with states that have fewer poor residents tending to have more college-educated residents. The glaring exception is Washington DC, which has high values for both variables, although it may be unfair to compare this city to larger states. We might also ask how much variation of the values there is for either variable within this limited geographic region. The sorted plot shows that there is substantial variation of these socioeconomic measures, with about a twofold range for each variable across the seventeen southern states. These types of questions are much easier to answer using the revised design. We will return to this example in a later chapter to see whether the patterns in southern states hold for the entire United States. The next sections provide examples of the three different types of micromaps.

1.3 LINKED MICROMAPS

Linked micromap plots add geographic context to the row-labeled plot design by including small maps in a new column of panels for each perceptual group of rows. Figure 1.3 shows the result of converting the row-labeled plot of Figure 1.2 into a linked micromap. Each dot that represents a state value is still horizontally aligned with its state name. Each state is represented by a polygon in the small maps, but these polygons are not aligned with their state names and values. For example, Delaware is below Maryland in sorted row order, but to the right of Maryland on the map. How can a fast link be established between polygons in a map, state names, and state values? The answer is to use color combined with perceptual grouping.

As shown in Figure 1.3, the linked micromap design adds a colored dot or other symbol to the left of a state name and uses the same color to link together all of the elements for each state. As we saw with row-labeled plots, separating the rows into perceptual groups helps the reader to focus on the values of a few mapped regions at once and to quickly identify clusters of mapped regions with similar values of the plot variables. With just a few states in each group, only a few distinct colors are needed to distinguish the states from each other within each group. Because these

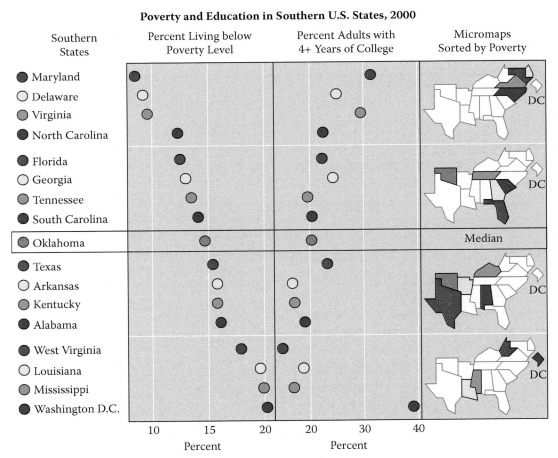

FIGURE 1.3 The linked micromap design adds the geographic context to the data, showing clusters of similar poverty and education in several portions of the southern region that are not apparent in Figure 1.2.

groups are visually independent, the same colors can be used for each group. The repeated use of the same colors from group to group can be confusing on first encounter. When two states in different perceptual groups have the same color, it only means that they have the same ranks within their respective perceptual groups and nothing more. Thus, for these simple plots, color is only used to identify the geographic location for each row's values and does not represent any level of mapped values, as it would on traditional choropleth (area-shaded) maps. The idea of interpreting the color simply as links within the scope of a perceptual group is easy to learn. We will explore the extended use of color in later chapters.

Adding geographic information to the plot shows clear geographic clusters in the socioeconomic data that we could not see using the simpler row-labeled plot designs—the Atlantic coastal states from Delaware to North Carolina tend to have the highest education and lowest poverty levels of this region, followed by a group of southeastern states (Florida to South Carolina). States in the western part of this region have levels lower than the median (Oklahoma). The last map shows that high poverty levels are found in West Virginia, Louisiana, and Mississippi and identifies the outlier as Washington DC, which has the highest proportion of college-educated residents but also the highest poverty rate in this census region. (Note that DC is shown larger and to the right of the main map for better visibility.)

In other applications, the mapped regions of linked micromaps might be comprised of standard administrative units such as states or counties in the United States, provinces in Canada, France, China, or India, or prefectures in Japan. Micromaps also work for subdivisions of these administrative units, such as counties within U.S. states. The mapped regions could be geographic units developed within a scientific community for more meaningful analysis. For example, environmental scientists in North America find many uses for the ecoregions developed by Omernik (1987) or Bailey (1995). The maps need not be for a standard geography at all, but could represent the location of retail outlets in a convention center or positions of baseball players on the field, as we will soon see.

Although we focus on micromaps in this book, there are certainly other graphical tools available. The association between two variables would traditionally be examined by a scatterplot. Replotting the education and poverty data in this format (Figure 1.4), we clearly see

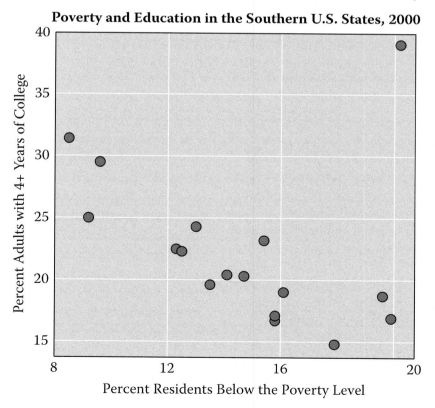

Poverty and Education in the Southern U.S. States, 2000

FIGURE 1.4   A scatterplot of the same college education and poverty data shows greater detail for the two variables' values and association but ignores the geographic information in the data.

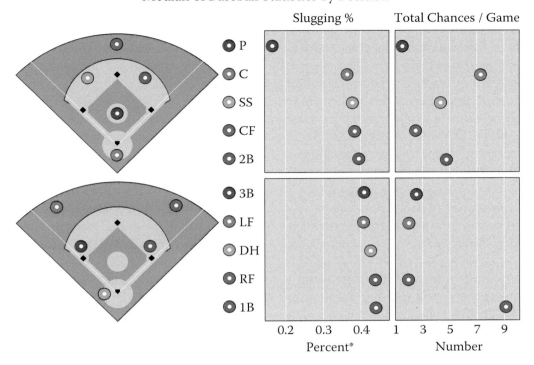

FIGURE 1.5 The micromap plot can represent any two-dimensional space, not just latitude-longitude on the earth's surface. This linked micromap displays the median of slugging percent and total fielding chances per game by position for major league baseball players during the 2007 season. "Location" is where the player usually stands. Key: P = pitcher, C = catcher, SS = shortstop, CF = center fielder, 2B = second baseman, 3B = third baseman, LF = left fielder, DH = designated hitter (bats but doesn't field), RF = right fielder, 1B = first baseman. *Slugging percent = (the total number of bases recorded by the batter)/(the total number of at bats); e.g., a home run counts for four bases, a triple counts for three bases, etc. Total chances per game = (the total number of putouts and assists credited to the fielder and errors charged to the fielder/number of innings played)*9 to scale to a full nine-inning game.

the expected negative association across the states, with the exception of a single outlier (DC), and the range variation is clear from the marginal range of each variable.

Scatterplots are the preferred medium for adding smooth curves to show a causal functional relationship or an association, although we will demonstrate later how to add smooth curves and to highlight departures from them in micromaps. However, despite the advantage of the scatterplot for seeing some types of patterns, the linked micromap design adds geographic location to the information displayed and so enables searches for geographic patterns that the scatterplot omits.

Our first micromap example (Figure 1.3) illustrates a typical state mapping application of the method. However, location in a micromap plot need not be represented in standard geographic units. Figure 1.5 is the same linked micromap design as Figure 1.3, but now the "geography" is a baseball diamond. The players' positions are represented by circles where they would

usually play. Performance statistics are for players in the U.S. major leagues during the 2007 season. The slugging percent, a batting average weighted by the number of bases a player got on each hit, is a measure of the player's offensive contribution. (Note that we follow convention in calling this a percent, although it is actually an index or ratio.) The total chances for putouts and assists per game are a measure of the player's defensive contribution. A more common fielding statistic, the percent of fielding errors committed out of the total number of opportunities, did not provide clear distinction among the positions, as all of the medians were under 5%. This is not surprising because of the selection bias of who plays each position—a player who makes many errors will not get to play regularly or will not stay in the major leagues for long. Medians of these statistics were calculated by position for players with at least fifty at bats or ten fielding chances (five for pitchers). Note that in the American League designated hitters only bat and so do not have a fielding statistic.

Tables of baseball statistics do not convey the pattern that is obvious in Figure 1.5, i.e., that players in positions up the middle of the field (not just the pitcher) have poorer slugging percents than those who play on the left or right side. Of course, some of the slugging percents are nearly tied and other statistics may better reflect the contribution of players to team offense. Comparison of the panels in Figure 1.5 suggests that players up the middle can be competitive through their contributions to team defense. That is, except for pitchers, who are generally poor batters, and first basemen, who naturally get more chances for putouts than the other players, there seems to be a negative correlation between these two statistics. This example illustrates the flexibility of the micromap design—any locations that can be represented in two dimensions can be displayed in this way.

1.4 CONDITIONED MICROMAPS

The primary purpose of the conditioned micromap is data exploration, unlike the linked micromap plot, which is used most often for presentation. A major difference between these designs is that linked micromaps use a single ranked variable to partition the mapped regions into a linear sequence of panels. In contrast, conditioned micromaps use two ranked variables to partition the regions into a two-way grid of panels, with the rank order of one variable determining the row membership and the rank order of the second variable determining the column membership. Two slider bars set category cutpoints for these variables, allowing the analyst to dynamically explore the geographic patterns on the map, based on categorized values of the two auxiliary variables. Another major difference is that conditioned micromaps use color to encode data values on the map while linked micromaps use color to link components across each row, without regard to the data values. Conditioned micromaps build on the choropleth map framework and inherit both the visual appeal and documented limitations of that map design (Dent 1993). The conditioning addresses the limitation of encoding just one variable and, in a dynamic setting, supplements rough area-of-color impressions by dynamic statistical feedback and alternative views. These and additional software features are described in Chapter 5.

A cancer control planner might want to know where mammography use is low and how the geographic patterns of screening are related to health insurance coverage and income. Let's see how a conditioned micromap could help to answer these questions. Figure 1.6 maps the state percents of women ages fifty and over who have had a mammogram in the previous two years, consistent with cancer prevention guidelines. These data are from the Behavioral Risk Factor Surveillance System survey, with percents averaged over the years when this question was asked from 1998 to 2006 (Centers for Disease Control and Prevention 2003). Normally we would use red to represent high numeric values, but here we use red to suggest a problem, i.e., low mammography rates. The percents are categorized into three groups with approximately 33% of all women ages fifty and over in each color-coded group (National Cancer Institute Cancer Statistics Branch 2003). This map shows that most of the states with high screening rates are along the U.S. coasts, whereas most of the low screening rates are in the Mountain and South Central states.

In Figure 1.7 the single map from Figure 1.6 is split into a three-by-three grid of nine maps according to levels of per capita income and health insurance coverage. The bars at the bottom and right of the grid are sliders that control interior thresholds to define the low, middle, and high classes for these two variables. States with low values for percent insurance are highlighted in the left column, and states with low values for per capita income are highlighted in the bottom row. The map in the lower left panel highlights ten states with low values for both insurance coverage and per capita income. Most states in this map are colored red, indicating a low screening rate. The population weighted average of the percent with mammograms for these ten states is 73.7%, as shown in the upper right corner of that map panel. This is lower than for any of the other maps.

Cancer control planners can interactively explore the geographic patterns on the map, conditioned on categories they create from variables they choose. If women have insurance or discretionary income with which to pay for mammograms, then a low screening rate is more likely due to inconvenient screening locations or a lack of awareness of the importance of screening. On the other hand, if low screening rates are mostly in areas with lower insurance coverage and lower income, then perhaps more free screening clinics are needed. Knowledge of these patterns can help planners to devise suitable programs to address the screening needs of women in different areas. It is clear from this conditioned look (Figure 1.7) that low screening rates (red) tend to be in states with lower income and lower health insurance coverage (bottom left map).

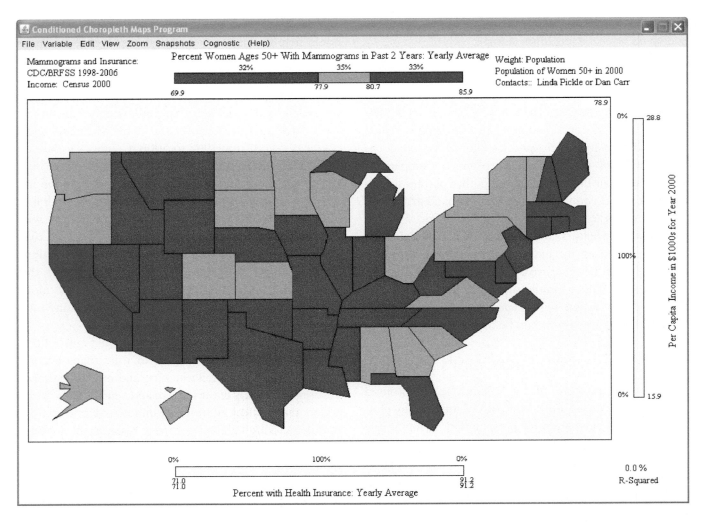

FIGURE 1.6 Full map of the percent of women ages 50 and over who have had a mammogram in the past two years, by state, which we partition into conditioned micromaps in Figure 1.7. **Note that red is used as the low (bad) value, contrary to traditional use.**

1.5 COMPARATIVE MICROMAPS

Comparative micromaps are one- and two-way sequences of maps indexed by time or other attributes. The emphasis is on comparisons across maps rather than comparisons within one map. The most common variant of comparative micromaps is a sequence of time-specific maps of values accompanied by a corresponding series of maps of the class or value differences of consecutive maps, allowing the reader to see explicitly the amount and location of changes. It is these difference maps that distinguish comparative micromaps from a standard time series of maps. The variables used in comparative sequences can be differences, ratios, or other computed values that are themselves inherently comparative, helping the reader to focus on the changes from map to map. There are many forms of comparative micromaps that can reveal patterns in data, but

most have features similar to those of linked and conditioned micromaps, such as simplified boundaries and other visual focusing methods.

To illustrate, Figure 1.8 maps the unemployment rate for U.S. states from 2001 to 2004. In order to color code the states in a consistent way across the yearly maps, all state unemployment rates for the four years (so, 51*4 = 204 values) were ranked. Then the highest 20% of these values were shaded red, the lowest 22% were shaded blue, and the rest were shaded gray. In the second row, each map has shaded only those states that changed categories from one year to the next. Rather than displaying differences in rates from one year to the next in this example, these maps simply show the new color for the states that changed. For example, New York changed from the middle (gray) category in 2001 to the highest (red) category in 2002, so the 2001–2002

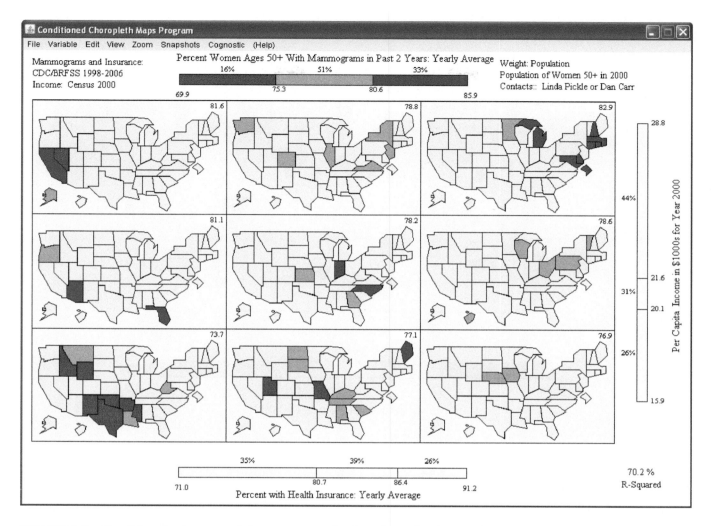

FIGURE 1.7 Conditioned micromap of the mammography data shown in Figure 1.6, conditioned on the per capita income (rows) and health insurance coverage (columns) as defined by the slider bars.

difference map shows New York in red, its newer color. New York stayed in the highest category from 2002 to 2003 and so is not shaded in the 2002–2003 difference map, but is shaded gray in the 2003–2004 map, indicating that it dropped back to the middle category in 2004.

This may seem overly complicated when we could just look at the first-row maps and see which states changed categories, but interpretation of these unemployment patterns should demonstrate the advantage of this design. As we can see from the first row of maps in Figure 1.8, there were more high-rate states (red) in 2002 and 2003, with persistently high rates on the West Coast. Rates in the North Central states were persistently low. Our eye may be drawn to the 2003 map because it has the most high-rate states, but examination of the difference maps in the second row shows that most states changed to the highest category between 2001 and 2002. Only

five more states joined the high category in 2003; then there was an improvement from 2003 to 2004.

These maps are consistent with what we know about the U.S. economy during this time. A series of economic shocks began in 2000—first the "dot com" bubble bursting in 2000, then the terrorist attacks in 2001, followed by several accounting scandals in 2001–2002. The stock market dropped sharply after each of these events and did not begin to recover until 2003. The national unemployment rate, which had been at what is considered a full employment level of 4.0% in 2000, rose throughout this recessionary period, peaking at 6.0% in 2003. As conditions improved, companies began hiring and the unemployment rate began to drop (5.5% in 2004). However, the comparative micromap plot demonstrates that not all regions were equally affected. States in New England, the Washington DC area, and the North Central region had among the lowest unemployment rates during

FIGURE 1.8   A typical comparative micromap plot design, with time series maps supplemented by explicit difference maps.

this period when rates in other states were rising, and relatively high unemployment persisted on the West Coast as other regions improved.

1.6   SUMMARY AND PREVIEW OF BOOK CHAPTERS

In this chapter we have introduced the concepts and basic types of micromaps with examples illustrating their wide variety of applications. At first glance, some of the examples in this book will appear to be simple, but further inspection may reveal complexities of design, purpose, and interpretation. Each design has strengths that make it a good choice for some tasks and some audiences, but all designs involve compromise and have weaknesses—there is no single design that will answer all questions for all types of readers. In our tour we strive to balance our enthusiasm for micromaps with comments about limitations, as awareness of limitations can lead to improvements. Some of you may well see solutions that we have missed.

This book is for both users and designers of statistical graphics. While some numerical literacy is important, we want to stress that micromaps are for people of all disciplines and are not limited to use by those

trained in statistics. Many of the elements of micromaps, such as confidence intervals, are routinely used in scientific publications, and statistical plots have been taught in grade school in recent years, so the material presented here should not be technically challenging. In addition, just as users of medical statistics are not expected to know all of the clinical processes that produced them, a micromap user is not expected to know the principles of the cognitive extraction of information from graphics. However, the micromap *designer* must be aware of these principles in order to design an effective graphic. We summarize these principles in Chapter 2, along with a brief history of research that led to the development of micromaps. In Chapter 3 we provide specific data visualization design guidelines based on the cognitive principles of Chapter 2. We then follow with chapters delving into each type of micromap: linked micromaps, conditioned micromaps, and comparative micromaps. Then in Chapter 7 we apply all three types of micromaps to demographic data for Louisiana before and after Hurricanes Katrina and Rita. We hope to challenge and inspire you to visualize your data in a new way so that you can communicate and explore data patterns more effectively.

2 Research Influencing Micromap Design

2.1 INTRODUCTION

The ideas for micromaps evolved from prior advances in statistical graphics but were influenced by research in cartography, cognitive psychology, computer science, image processing, and even graphic arts. We continue our tour by summarizing the most influential of these earlier developments. If you are eager to see more micromaps you are free to jump to Chapters 4 to 7. If you are interested in why designs are the way they are, please continue reading.

Both tasks and tools matter as we seek to find and understand patterns in data. In our efforts to weave together insights and methods from multiple disciplines, we noted that different disciplines historically focused on different tasks. Much of a statistician's training, especially in thinking about patterns, is related to the statistical tasks of describing and comparing distributions and to creating and refining models that describe how variables are related. There is little direct focus on the tasks of pattern identification, distribution comparison, and model building in the web page design and usability literature. Instead, that community is more focused on searching for and filtering information, drilling down to find a specific piece of information and navigation on the web. Nonetheless, good tools for one purpose often can be adapted to another purpose. In recent years subcommunities have emerged in nonstatistical disciplines that have a strong interest in data analysis, leading to interest in analytic tasks and tools that more greatly overlap those of the statistical graphics community.

Since people vary in visual, cognitive, and analytic skills, we need to know what skills are common to us all and how we vary so that we can design tools that better fit the range of human performance. Unfortunately, it has been easier to produce tools than to conduct studies of their use that control for our many differences. This has led to conflicting study results. For example, we found little difference in statistical map-reading performance between experts and novices (Lewandowsky et al. 1995; Maher 1995). In contrast, a recent study by Ware and Mitchell (2008) found that skilled observers were better at seeing patterns in complex three-dimensional network graphs than unskilled

subjects. They also found that observers could see much more detail and interpret the graphs more accurately using the study's very high-resolution displays with a volume- and lighting-model-based rendering of three-dimensional lines (i.e., tubes). In other words, previous assessments of human performance on similar tasks reflected human capabilities as constrained by the visualization environment available at that time.

In this chapter we summarize the research that directly impacted the development of micromaps and then turn our attention to the ongoing research in human perception and cognition that guides our visualization designs today. This information is critical for aspiring designers who wish to make their visual tools easy to use and accurate in conveying the underlying statistical data. We hope that even experienced designers will find a useful insight or two as they think about possible refinements to our designs. Cognitive scientists will see that more guidance is needed for data visualizations with multiple and often conflicting objectives.

2.2 INFLUENCE OF STATISTICAL GRAPHICS RESEARCH ON MICROMAP DESIGNS

Much has been learned about how readers interpret statistical graphics over the past few decades, as quantitative graphical software has become available to nearly everyone with a computer. As will be seen in Chapter 3, micromap designs strive to incorporate modern guidance from quantitative graphics designers and researchers in human perception and cognition, while drawing from such varied disciplines as visual perception, image processing, cognition, geovisualization and cartography, and human-computer interaction. Until recently, the various research areas were studied independently with little cross-fertilization of ideas and few outlets for cross-disciplinary publications. The recent confluence of these disparate disciplines led to the development of micromaps. While this development was partially inspired by the statistical graphics community in the 1980s, we limit our discussion here to topics directly related to micromap designs. Readers who wish to delve into more of the historical research related to micromaps are referred to Dan's website (http://mason.gmu.edu/~dcarr/). Michael Friendly has prepared several extensive histories of data visualization in general, which are available at his website and in print (Friendly 2008; Friendly and Denis 2001).

Although statistical graphics date back over 200 years (Playfair 1786), John Tukey revolutionized exploratory statistical graphics in the 1970s with his introduction of intuitive ways to represent data distributions and comparisons (e.g., stem-and-leaf plot and box plot) (Tukey 1977). Other statisticians expanded on this work during the next 20 years, notably William Cleveland and Edward Tufte. Cleveland demonstrated by cognitive experiments that readers were better able to judge values that were represented as position along an axis than those represented by other encodings (Cleveland 1985; Cleveland and McGill 1984). In linked micromaps values are read using position along a scale, consistent with this finding. Tufte focused on effective presentation and recommended a simple graphics style, devoid of "chart junk" (Tufte 1983, 1990). He also argues for small multiples for comparison, an economical display method since the reader only needs to learn the meaning of the graphical design once and then can focus on the data (Tufte 1983). Micromap designs follow these principles by arranging small multiples of maps in an organized way.

Advances in computing and graphics hardware and computing languages in the 1970s and 1980s widened the door for the development of statistical software for interactive and dynamic graphics. Data exploration and graphical analysis research topics included multivariate analysis, human computer interface tools, large data set analysis, and data analysis computing environments. Many micromap interactive and dynamic features connect to research that occurred during this period. For example, linking across panels by color and interactive data selection became commonplace. The conditioned and comparative micromap designs adapt some of the methods used for early dynamic computing and visualization environments.

Much of the early statistical graphics research concerned studying the variation in data using methods that went far beyond the 1970s' business graphics of bar charts, pie charts, and time series line charts. The scatterplot, most commonly used to view the scatter of data, also supported viewing two-dimensional projections from higher-dimensional data, such as from rotating three-dimensional views or grand tour sequences of projections. The scatterplot inherently uses the range of variables to set the limits of axis scales and most often uses linear scales. This provided an initial scale for focusing on the data in order to study its variation, which is a key theme in micromaps.

Focusing methods are important for micromaps. That is, the reader's attention can be focused by selecting study units in one view and by highlighting them in all relevant available views. The other study units remain in the background to provide context for interpretation. There are instances when providing context motivates scale modifications to include zero or other reference values that are outside the range of the data. When seeing variation in the data is important for the task at hand, changing the scale to include zero reduces the area available to show variation in the data, and so is not recommended. In most micromaps we assess the variation of highlighted study units relative to the variation of all the study units or relative to specific subsets of study units.

Advances in the late 1970s through the 1980s included interactive and dynamic methods for selecting study units to highlight, coupling this with rapid display of all the study units. These methods included point-in-polygon selection, painting, and linked brushing (Becker, Cleveland, and Weil 1988; McDonald 1982; Newton 1978). An early map application applied point-in-polygon selection to a latitude and longitude coordinate plot of acid deposition monitoring locations (Carr et al. 1987). This example compared the distributions of deposition variables in a scatterplot matrix and showed sample time differences in a comparative dot plot.

Micromap graphics differ from most of the above methodology in two ways. First, by definition, micromaps always include maps among the views of study units. Second, micromaps use different methods to highlight study units. Linked micromaps sort the study units, partition them into small subsets, and systematically highlight these subsets. The conditioned micromaps and many comparative micromaps use a three-class slider to partition study units into three classes as a basis for highlighting. Carr and Nicholson illustrated the use of a three-class slider for dynamic four-dimensional hyperplane slicing that masked points falling outside of the acceptance interval (Carr and Nicholson 1988).

An important focusing method is **conditioning**. This selects study units for emphasis based on whether conditioning variable values fall into specified intervals or classes. This is not the same as **filtering**, which shows only the study units that meet the selection criteria. In contrast, conditioning partitions the data into multiple views that each contain highlighted study units so that

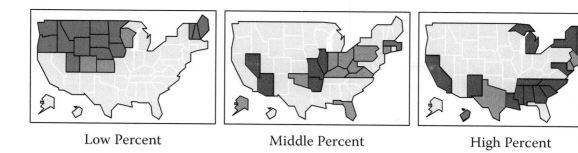

| Low Percent | Middle Percent | High Percent |

FIGURE 2.1   An example of one-way conditioned micromaps. The study variable has low, middle, and high values encoded as blue, gray, and red, respectively. Conditioning partitions the states into three panels: states with low, middle, and high percents are highlighted in the left, middle, and right panels, respectively.

we can systematically compare and contrast the subsets. In Figure 2.1, blue states dominate the left panel and red states dominate the right panel, evidence of a positive correlation between the study and conditioning variables. A scatterplot would show the relationship between the two variables in more detail, but would not convey the spatial patterns shown in these micromap panels. Using conditioning to define a comparative grid of panels, as in this example, changes an investigation from a sequential filtering of one variable at a time to more of a multivariable approach. In this context we can assess functional relationships, densities, or geospatial patterns within panels as well as changes across panels.

As computing power increased, so did the typical size of data sets. An increasing number of observations leads to overplotting problems, making the visualization less effective. Carr's papers on **hexagon binning** are relevant to micromap design for large data sets (Carr et al. 1986; 1987; Carr 1991). As you can see in Figure 2.2, heavy overplotting in the scatterplot of 10,000 points masks any underlying pattern. Aggregating the data and

displaying the number of observations in each hexagon cell, either by symbol size or color, shows the circular pattern expected of the bivariate, independent and identically distributed normal random data. Hexagon binning, gray-level erosion, and bivariate box plots described in the 1991 paper were subsequently used for linked micromaps in a data mining context (Carr et al. 1997). The direct display of two-way indexed bivariate box plots and their differences was a direct precursor to two-way comparative micromaps.

The 1987 hexagon binning paper also illustrated the **direct display of differences** between two hexagon binned scatterplot matrices using symbol size and color. Using color to indicate an overplotting of points from different subsets in a time plot showed that the monitoring of eastern sites started earlier than that of western sites, indicating that the data were not totally comparable in time. The juxtaposed direct display of differences between plots or maps is an important theme in this book and is emphasized in the chapter on comparative micromaps.

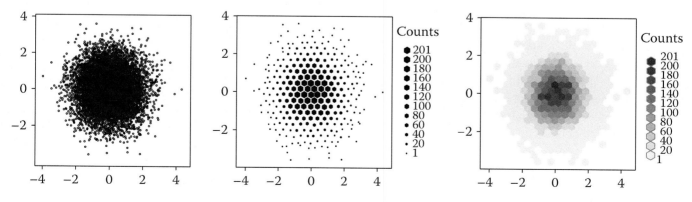

FIGURE 2.2   An illustration of hexagon bin plots displaying 10,000 pairs of independent normally distributed random values. The leftmost plot is the usual scatterplot, with too much overplotting to see the circular density of the data. The center and rightmost plots overlay the plot area with a hexagon grid and display the number of pairs observed in each grid cell by symbol size (center) or color intensity (right), making the pattern clear.

In addition to visualization problems, larger data sets also result in a greater number of subsets possible for viewing, conditioning, and comparison and more combinations of variables to consider. John Tukey proposed the idea of computing **cognostics**, an automated ranking of plots by some statistical measure for subsequent analyst review (Tukey 1988a). Inspired by this concept and a presentation by Paul Tukey on scatterplot diagnostics (scagnostics), Carr addressed the issue of too many plots to review in a streaming parallel-computing environment by saving the data necessary to produce the most interesting plots for subsequent review (Carr 1991). The cognostics idea is applied to dynamically conditioned micromaps in Chapter 5 where an algorithm ranks a subset of the billions of possible combinations of slider values in order to suggest some good settings. Other methods have been proposed over the years to identify meaningful data views, such as projection pursuit (Friedman and Tukey 1974). In recent times Wilkinson and colleagues have rekindled interest in scagnostics (Wilkinson, Anand, and Grossman 2005; Wilkinson and Wills 2008).

There were exciting developments of graphics software and **data analysis environments** in the 1980s. Visualization designers in the 1990s began to utilize the knowledge compiled by these early groups to develop new types of graphical representations and to study the effectiveness of existing ones. Rapidly improving, more powerful, and widely available computing platforms led to the development of the more highly interactive systems for visual data exploration and analysis, such as XGobi, a descendant of DataViewer that combined high-dimensional projections, linked scatterplot brushing, and matrices of conditional plots (Swayne, Cook, and Buja 1998; Buja, Cook, and Swayne 1996). At the end of the decade, Wilkinson published the object-oriented grammar of graphics that puts the production of graphics into a mathematical structure (Wilkinson 1999, 2005). This formalism leads to an understanding of equivalences and relationships between types of graphics and to new combinations of features.

Some developers have broadened the functionality of these early systems over time to produce a stand-alone graphics visualization program (Swayne et al. 2003; TIBCO Software, Inc. 2008) or one linked to the analytic features of other commercial software packages (Cook and Swayne 2007; Symanzik et al. 2000; Wilkinson 1984). Symanzik's work specifically links statistical graphics to geographic information system (GIS) mapping software. Other developers, such as

MacEachren's GEOVISTA group at Penn State, have taken a different approach and have instead constructed a developer's tool kit of interchangeable objects and modules by which others can create customized graphical programs (Takatsuka and Gahegan 2002; Weaver 2004; Weaver et al. 2007). This approach is similar to software development in Java™ (Sun Microsystems), where modular "beans" can be combined to customize program functionality. By the end of the 1990s, many of these visual tools were adopted by applications specialists and began to appear in routine publications.

2.3 CONTRIBUTIONS FROM OTHER RESEARCH AREAS

Methods of communicating information for the purpose of informing, educating, or persuading target audiences have been studied for many years. Models of human communication have evolved from a basic two-step message delivery, i.e., developing a message and then delivering it. Now more complex models first define the target audience and then motivate them to act in order to receive a message delivered through a particular media channel (Littlejohn and Koss 2007). Effective communication is more difficult when the information to be communicated is complex and not familiar to the target audience. Health communicators, for example, have attempted to summarize scientific discoveries into understandable messages and to develop effective delivery methods in order to influence policy and to encourage healthier behaviors. However, it is only recently that researchers in this field have brought their expertise to bear on the communication of quantitative information to the lay public (Nelson, Hesse, and Croyle 2009). Their evidence-based communication recommendations are based on research in many of the same disciplines that influence statistical data visualization.

Computer science research and development, from database methodology to color options that we take for granted, has influenced us all. We limit our discussion here to the more focused computer science area of human-computer interaction research that most impacted micromaps. This area introduced many ideas for improved data visualization, reflecting a strong interest in harnessing increasing computing power, advances in graphics hardware, and new computer interfaces (Card, Mackinlay, and Shneiderman 1999; Shneiderman 1992). Some of the dynamic querying tools studied were radio buttons, tree maps, and slider bars. There were other computer scientists who worked with statisticians to produce systems and contribute ideas. For example, the four-dimensional system that Dan used was built by computer scientists who provided both hardware engineering and underlying software tools that supported the visualization. Research in computer science initially had limited influence on statistical graphics because realistic rendering and dominant topics of that time did not address comparisons of data or other statistical questions important for data analysis. However, over time these visualization methods became important in other areas, such as GIS and data mining. The sophisticated use of color graphics for more realistic imaging and more recent advances for making movies are slowly becoming incorporated into statistical graphics software. Now there is more sharing of methods because of the increased cross-disciplinary interest in quantitative data and because of commonly available computing platforms.

Naturally, the area of **cartography** has influenced micromaps through its recommendations for map designs. Classed choropleth maps, used in the micromap plots, shade each geographic region by a color representing a range of values of the variable being mapped. Cartographers recommend assigning colors according to the visual metaphor, i.e., darker shades representing higher values (Dent 1993). Alternatively, a diverging color scheme can be used that emphasizes the extremes of the range of values. Common methods of assigning values to categories are equal length intervals, e.g., each color represents a range of five units, or by equal numbers of geographic regions in each category (the quantile method). The former is suitable when the values being mapped have absolute meaning, such as the number of packs of cigarettes smoked per day, or when the actual mapped values are of interest, but the latter is preferred when rank order is more important. The categorization method is an important design choice requiring consideration of the pros and cons of each method. For example, the equal width intervals can force most of the values into the middle categories in order to encompass extreme values (outliers), and the quantile method forces the specified fraction of geographic units into each category, possibly distorting the visual impression of the data distribution. For the sample data mapped in Figure 2.3, only a few states have rates in the extreme categories on the top map, but ten states are forced into each of these categories on the bottom, giving the impression of greater geographic variation on the bottom than on the top. The same data are displayed on both maps, but the impression of the geographic pattern is different depending on the categorization method.

We discuss color selection and category definitions in more detail in Section 2.4.4.

Given the small size of micromaps, the blocks of color on choropleth maps have the advantage of being more visible than if the values were displayed by small symbols or hatch patterns on the map. Using highly saturated colors makes small areas stand out even more (Brewer 2005). On the other hand, the eye can be drawn to large blocks of color that represent small populations, e.g., Wyoming versus the more populous Washington DC, distorting the visual impression of the geographic distribution of people affected by the disease, for example. A micromap re-design may attempt to mitigate this areal bias by increasing the size of small U.S. states, but the analyst needs to be aware of this potential problem when using micromaps to communicate to others. The conditioned micromap design can partially address this issue by conditioning on population.

Alan MacEachren (1995), in his book *How Maps Work*, was one of the first to consider map design through the lens of visual cognition and perception, considering maps as a subset of symbolic knowledge representation. This approach was used to conduct studies of how readers extract information from statistical rate maps at the National Center for Health Statistics to inform the design of a new mortality atlas (Pickle et al. 1996; Pickle and Herrmann 1995). Early focus groups with epidemiologists and other public health professionals, the primary user group for the atlas, identified three types of questions that they wanted to answer, closely following Bertin's classification of general visualization tasks (Bertin 1973):

- Rate readout task: What is the disease rate in a certain place?
- General pattern recognition task: Are there geographic trends in the data or clusters of high- or low-rate areas?
- Map comparison task: Are the rate patterns similar for males and females, or for blacks and whites?

Inconsistency of results by cartographic researchers in the past is due in part to drawing conclusions with only one of these questions in mind. Subsequent cognitive experiments found that choropleth (area-shaded) maps with standard vertical legends, as shown in Figure 2.3, were preferred and used most accurately by novice map readers (Pickle, Herrmann, and Wilson 1995). Blanking out areas with unreliable rates (compared to a hatched

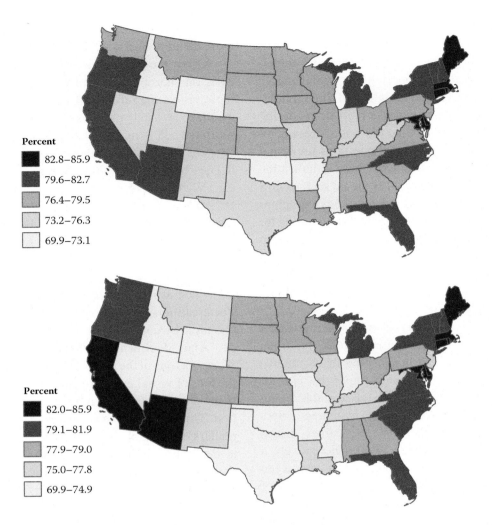

FIGURE 2.3 Comparison of categorizing the same values by equal width intervals (top) and quantiles (equal numbers of states in each interval bottom). There appear to be greater geographic differences on the bottom map.

overlay or separate reliability map) was found to interfere with pattern recognition tasks (Lewandowsky et al. 1993; MacEachren, Brewer, and Pickle 1998). In addition, classification of standardized rates by quantiles led to fewer map-reading errors than other popular categorization methods (Brewer and Pickle 2002). These findings were recently used to critique the designs of all U.S. mortality atlases (Pickle 2009).

Data analysts often devised visualization designs that were later found to be compatible with our cognitive strengths and weaknesses. That is, sometimes our intuitive designs were later theoretically justified. For example, some in the statistical graphics community recommended displaying differences explicitly even before they learned about the cognitive difficulties inherent in image comparisons. On the other hand, most early cognitive researchers were narrowly focused and did not address issues important for data visualization.

Research on visual perception, what and how we see, and cognitive perception, how this is translated and remembered in the brain, also has a long history, which will only be briefly summarized here. We will delve more into the details of our perceptual strengths and weaknesses in the next section.

It wasn't until the 1980s that visualization developers began to incorporate the results of **cognitive research** into their designs. Cleveland's work on cognitive extraction of information had the most impact because it was directed at statisticians. Tufte proposed displaying multiple small graphics for direct comparison, rather than relying on recall of serially presented images (e.g., animation), an idea later recognized to be consistent with our short-term memory limitations for complex images (Tufte 1983).

Cognitive scientists began to apply their methods to the study of graphs and maps during the 1990s. Stephen

Kosslyn described how the brain processes images in *Image and Brain* (Kosslyn 1994b), part of the new field of cognitive neuroscience, and applied these theories to practical graph design (Kosslyn 1994a, 2006). Kosslyn's book (1994a), recently expanded (2006), was very helpful in several areas of Dan's early micromap designs, especially by refining and strengthening his notions about perceptual grouping and providing pointers to the cognitive literature. For a broad view of human vision and perception, see Palmer's book *Vision Science: Photons to Phenomenology* (Palmer 1999). In a later book, Colin Ware brought together research in visual perception, color theory, visual attention, memory, and space perception to propose design guidelines for both static and animated data visualizations (Ware 2004). The most recent book in this area is Stephen Few's *Now You See It* (2009), which applies these principles to numerous types of graphical analysis.

Following these beginning efforts to integrate visualization research from statistics, cartography, and cognitive and computer sciences, the term *visual analytics* was coined recently to describe the science of analytical reasoning facilitated by highly interactive visual interfaces (Thomas 2007). A subset of data visualization, this cross-disciplinary area expands beyond traditional information visualization to include knowledge representation, management and discovery techniques, decision sciences, and cognitive and perceptual sciences. Recommendations for future research include seeking to better understand how people use visual methods to synthesize quantitative and nonquantitative information in order to gain insight about patterns in the data, to make timely decisions, and to communicate results so that appropriate action can be taken (Thomas and Cook 2005).

2.4 HUMAN PERCEPTUAL AND COGNITIVE STRENGTHS AND LIMITATIONS IMPACTING DATA VISUALIZATION

People have different approaches to reasoning about data, depending on their skills and experience, but research has shown that there are commonalities in their processing steps. Some researchers call this *sense making*. A classical statistical analysis is usually straightforward, consisting of sequential steps of experimental design, the conduct of the experiment, and a statistical summary of results. An exploratory analysis is often interactive and less structured. Usually there is a phase of information gathering and preliminary processing, followed by choice of the representation method that will address the question at hand or questions raised by preliminary graphics. Then the analyst gains insight by manipulation of this data representation, e.g., by choosing the variables to display on a linked micromap plot, exploring the data, asking questions and forming hypotheses, and then drawing a conclusion or making a decision (Bhowmick et al. 2008). The designer's role in these steps is to offer enough interface options to satisfy the analyst's needs and to make them as easy and intuitive as possible.

Users of the new exploratory visualizations have varying levels of technical and domain expertise and have been found to use somewhat different strategies and tools for data exploration (Bhowmick et al. 2008; Weaver et al. 2007). Thus, any new computer-based visualization systems should include interactive tools that enable data exploration at both the novice and the expert level, perhaps with a tutorial or wizard to guide inexperienced or first-time users. Keim's article on visual data mining provides an excellent review of these techniques (Keim 2002). Future research in the new area of visual analytics should lead to improved visualization designs, but for now we must rely on recommendations from the disparate scientific groups who have conducted experiments in the types of visualizations being considered.

As visual tools have grown more sophisticated, interfaces better match our sensory capabilities, and more powerful computers can process massive amounts of data, it has become apparent that the human mind's ability to detect patterns is greater than once thought. This is particularly true if the data are presented using visual tools that are optimum for the task, the user, and the type of data. Thus, by tapping into our perceptual and cognitive strengths, visual data exploration can often identify data patterns more quickly than traditional statistical or computational methods. A key problem of visual analytics, though, is the limitation of human short-term memory. Recent efforts at improving visualization designs have attempted to enhance the viewer's cognitive abilities by representing a large amount of data in a small space, organizing these data by space and time or other important variables, simplifying the design so that patterns are quickly recognized visually without additional thought (i.e., they hit us between the eyes), and including user aids such as gridlines and perceptual grouping of graphic elements (Card, Mackinlay, and Shneiderman 1999). In the next section we describe how we see and process visual data representations and summarize our perceptual strengths and limitations that impact their design.

2.4.1 A Model of Cognitive Processing

Psychologists have proposed numerous models of cognitive processing, i.e., how we think. While our thought processes are really too complex to be represented by a simple flowchart, it is helpful for visualization design to consider a typical sequence of three levels of cognition. Herrmann et al. (2006) provide an excellent review of this area, which we briefly summarize in this section. Heuer (1999) also presents a brief and very readable summary of cognition related to gathering and evaluating evidence, and Stafford and Webb (2004) provide experiments by which you can test your own cognition.

When presented with something new, we first perceive the information. This is a set of lower, more basic processes that are performed by specific areas of the brain. Then we comprehend, learn, and store what we learned in memory, sometimes with the aid of past memories. Finally, the brain's high-level systems help us to integrate different cognitive processes to reason, find solutions, make a decision, or communicate. In general, these three types of processes are progressively more difficult, and therefore take increasing amounts of time to perform. Thus, any aspect of using a data visualization that can be handled by a lower level of cognition should result in a faster time to response. For example, identifying a red symbol on a graph that represents a very high value is faster than matching a gray-shaded symbol to a legend, reading the value, and deciding whether it is very high (i.e., seeing is faster than thinking). To illustrate, can you find the word *eureka* in the left panel of Figure 2.4 (horizontal, vertical, or diagonal)? This is difficult and will take some time because you need to scan all of the letters. However, the red letters in the right panel stand out, drawing our eye immediately.

Within this general cognitive framework, stage theory, proposed by Sternberg (1975), hypothesizes that a series of information processing stages occurs between sensation and final response for any given task. We illustrate the importance of this concept with an experiment in statistical map reading. Herrmann and Pickle (1996) found that readers typically extract information from statistical maps by performing five separate stages: orienting to the map, reading the legend, integrating map and legend, extracting relevant information from the map, and drawing a final conclusion. The independence of these stages was supported by timings of combinations of map-reading tasks; e.g., the time taken by study subjects to orient to a map and then understand its legend was nearly equal to the sum of times required when the two tasks were done separately. This finding implied that the printed map page could be designed and tested one map element at a time (e.g., legend, map, title, footnotes), an easier task than designing an experiment to test the entire "gestalt" of the page. A similar series of stages was found during usability studies of web-based linked micromaps. We examine the general cognitive stages of sensation and perception, comprehension and learning, and retention and remembering in the next sections, pointing out their importance for visualization design.

2.4.2 Sensation and Perception

Seeing starts with the sensation of characteristics of an item, such as its color, texture, and shape (preattentive vision). Vision deficiencies, such as color blindness, will affect the accuracy of this step, as will the resolution of the image displayed on the computer monitor or the printed page. Perception is the translation of these sensations to something meaningful. We generally do this unconsciously by comparing the sensations to similar images stored in our memories. Our expectations and experiences affect how quickly and accurately this is done. For example, a person from a developed country could quickly recognize a drawing of an automobile even if most of the image is hidden, but a person from a remote place might not have "reference" images of automobiles in memory and so would have more trouble identifying the drawing. We have the ability to construct

```
fasexfasdfasdfkasdlfkajsdfdqofru
asdkfuasgqoftasdkfnasdfasdeeiode
jsdfkaslrkfasngaskdjfqrxpordrwks
jaonxckgaemvakdjfalskdbbbkxjfals
kjalslkfkesnfkdnouamcupoiwfrbua
rdfkdjkadffasdfasopd;fk,cfdaskdfas
```

```
fasexfasdfasdfkasdlfkajsdfdqofru
asdkfuasgqoftasdkfnasdfasdeeiode
jsdfkaslrkfasngaskdjfqrxpordrwks
jaonxckgaemvakdjfalskdbbbkxjfals
kjalslkfkesnfkdnouamcupoiwfrbua
rdfkdjkadffasdfasopd;fk,cfdaskdfas
```

FIGURE 2.4   Illustration of how a pattern is more quickly perceived when highlighted by a distinct color.

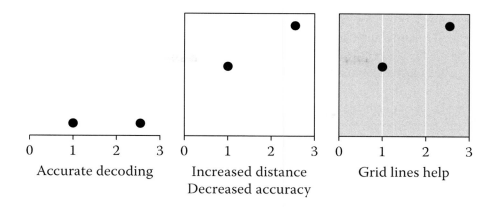

FIGURE 2.5   Encoding by position along a common scale enables high perceptual accuracy of extraction.

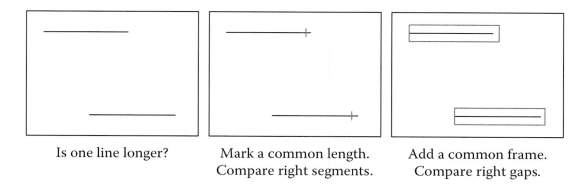

FIGURE 2.6   Judging lengths of nonaligned lines.

mental images even without seeing the object; psychologists call this mental imagery (Pylyshyn 1973). These images can be rotated and otherwise manipulated just as if they had been directly perceived, although they may not be remembered accurately (Kosslyn 1995).

Cleveland and McGill's experiments demonstrated that judgment of quantitative information from a graph was most accurate when the quantity was encoded as position along a common scale, followed very closely in accuracy by position along a nonaligned scale, as for two vertical bars not aligned to the same baseline (Cleveland and McGill 1984, 79). Other encodings, such as length, angle, area, volume, curvature, and color, were progressively less accurate. Cleveland proposed framed rectangles, i.e., a filled bar with a fixed-size outer frame, as a visual cue to help the reader to judge length along nonaligned scales (Cleveland and McGill 1984, 79).

To illustrate these principles, the left panel of Figure 2.5 shows that decoding values encoded using position along a scale is similar to reading distances using a ruler. The middle panel of the figure suggests that, not surprisingly, the farther away the ruler is from the points, the less accurate the decoding. The right panel illustrates how grid lines can help guide the eye to align the points with the ruler (axis) labels, in effect moving the scale up to the points. Judging lengths is more difficult, especially when the lines are not aligned at the same baseline, as shown in Figure 2.6. Visual length comparisons are based on ratios. In the panels, the ratio of bottom-line to top-line lengths is 1.02, hard to detect. Adding a visual cue such as a vertical mark or a common-sized rectangle framing each bar helps us to see the difference in lengths because we can focus on the lengths of the rightmost segments in the middle panel and gaps in the right panel (a ratio of 1.4).

2.4.3   ATTENTION

In order to perceive a visual stimulus, e.g., a patch of color on a map, we must focus our attention on the appropriate part of the image. Three types of blindness are present in people with perfectly normal vision. Among early perceptual psychologists drawing attention to these topics was Julian Hochberg, whose interest was in applying this knowledge to build motion picture scenes from a sequence of nonoverlapping views (Mark 2007). If our attention is focused elsewhere or we are not concentrating to a sufficient degree, it is surprising what we will not see—even a gorilla walking through a crowd (Simons and Chabris 1999). This is called **inattentional blindness**. On the other hand, our attention is drawn quickly (in about one hundred milliseconds) by movement, bright colors, or large objects. We can take advantage of this almost unconscious shift of attention in graphics design (besides creating pop-up Internet ads that draw your attention with dancing hamsters). This term may be new to you, but the phenomenon is behind recent concerns about driving while talking on a cell phone.

If two similar images are flashed rapidly before our eyes in an alternating sequence, the part of the image that is different will appear to blink, making it easy to find. However, inserting a blank image between them in the sequence will erase the first image from our memory so that we no longer can see the differences (Rensink, O'Regan, and Clark 1997). (For examples of this, see Rensink's website: http://www.psych.ubc.ca/~rensink/flicker/download/) This is called **change blindness**. If our attention is drawn to one local change in an image, then a second change that occurs within two hundred to five hundred milliseconds will not be seen, i.e., we have an **attentional blink**. Curiously, we could see the second change if it appeared more quickly in succession, less than one hundred milliseconds after the first. Experiments have shown that training such as meditation can reduce the attentional blink (Biello 2007).

These attentional deficiencies impact the effectiveness of animated or dynamic graphics. However, change blindness can also occur when our eyes jump from one part of an image to another, say between side-by-side graphics, because the shift of focus of attention takes longer than our brain can retain the first image. The graphics designer needs to help the reader overcome this visual deficiency. Possible ways to do this include showing all time-specific images rather than animating them, showing differences explicitly, or placing images that are to be compared as close together as possible. Cleveland's research showed that performance is worse the farther apart the images are, but we really don't know how close is close enough to prevent this effect (Cleveland 1985).

We are also limited in the number of items that we can attend to at once, called the span of attention— probably only three or four. This limitation probably explains why many graphic animations do not work— there is too much (more than four items) to attend to at once. The width of our field of attention, i.e., how much

we can see without shifting our attention, has also been found to be limited, with the limiting angle dependent on the degree of concentration required for the task and on other environmental factors. Training and repetition can improve our speed and ability to attend to specific details in a scene or to block out visual distractions (a good example is how a teenager's ability to focus and skill at a computer game improves after he or she plays it all day).

2.4.4 COLOR

The use of color is so fundamental in visualization design that its perception requires an in-depth discussion here. Using color well is not easy. Color is one of those concepts that everyone thinks they understand, but that is really more complex than it first appears. Since two people can have very different mental images of what "blue" looks like, there are several methods of defining precise colors. All are three-dimensional systems. Colors on computer and television screens are defined by combinations of red, green, and blue (RGB), but paint and printing ink are defined by mixing cyan, magenta, and yellow (CMYK, where K denotes black). The most precise definitions are from CIELAB (CIE 1932). Any of these could be used to define colors, but we have found that the hue/saturation/value (HSV) system is convenient for discussing visual designs. Brewer's book on map design and her recent paper on cancer map design contain more detail on color perception and choice, concepts that hold for graphs as well as maps (Brewer 2005, 2006).

Hue is the color dimension that is associated with wavelength of light and with names of colors, such as red, yellow, and blue. Most languages around the world include words for black, white, red, green, yellow, blue, brown, pink, purple, orange, and gray. Differences in hue are best used for encoding different attributes, as in a qualitative graph or unordered variables. Different wavelengths have different focal lengths, so what we "see" is a compromise between the actual and perceived distance to the image. Most people perceive long-wavelength colors, such as red and orange, as being closer to their eyes than short-wavelength colors, such as blue and green (Ware 2004). (Of course, the colors we see on a computer monitor or printed page are mixtures of hues, so our eye probably compromises, balancing focal length between the colors.) We can take advantage of these focal length differences if we wish to have part of our visualization appear to be in the foreground or background. For example, if there are red and blue lines that cross on a graph, the red line is most likely to be perceived as being plotted on top, whether it is or not. If we plot red lines on top, people will be more likely to judge line order correctly. If we wish to reduce, rather than emphasize, the difference between perceived distances to two hues, we can use colors that are in the red and blue family but are not pure hues. We have had success using red and blue-green together—the blue-green has a longer wavelength than pure blue and so does not appear to be as far behind the red graphic elements. In addition, blue-green avoids the red-green color blindness problem by moving toward blue from a pure green.

Saturation, also referred to as chroma or intensity, measures the purity of the color. A highly saturated color has little or no gray in it, while a highly desaturated color is almost gray, with none of the original color. You may be more familiar with the term *shade*, which refers to a mix of pigment and black paint, or *tint*, a mix of pigment and white paint. We only perceive a few different steps of varying saturation, so changing saturation alone is not effective for encoding a quantitative variable. However, the eye is drawn to highly saturated colors, so these can be used to good effect for drawing attention to a part of the visualization. In addition, highly saturated colors stand out more and so can be used as fill colors to improve the visibility of small symbols or areas. They are also good for encoding a qualitative variable, i.e., where there is no natural order to the categories. Tufte echoes cartographers' recommendations to use saturated colors sparingly and then only to highlight limited areas (Imhof 1982; Tufte 1990).

Value, also referred to as lightness, measures how much light seems to reflect from an object compared to the reflection from a white surface. Equal changes in lightness from light to dark are most easily and commonly used to encode values that vary from low to high, as shown in Figure 2.7. Brewer recommends combining changes in both lightness and saturation for map colors, as these will be more distinguishable one from another than colors that vary only by lightness or only by saturation (Brewer 2005). For example, choropleth map colors might range from light-desaturated red (light pink) to dark-saturated red (dark red).

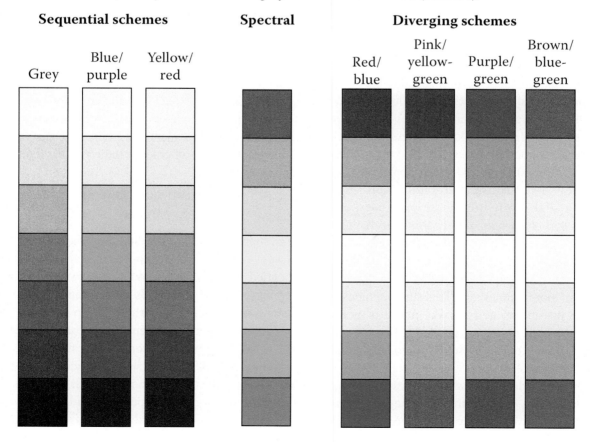

FIGURE 2.7 Sequential, spectral, and diverging color schemes tested for effectiveness in choropleth maps by Brewer, MacEachren, and Pickle (1997).

As part of a series of cognitive experiments in statistical map reading, color choices for maps were tested for effectiveness (Pickle and Herrmann 1995). Results are equally applicable to other types of visualization designs. Consistent with cartographers' prior recommendations, very distinct colors, i.e., different hues, were found to be best for reading values (or ranges of values) directly from the map (Hastie et al. 1996) while a sequential color gradient (light to dark) was best for pattern recognition tasks (Lewandowsky et al. 1993). A diverging color scheme, consisting of a gradient of equal changes of lightness/saturation within each of the two hues, combines the best features of both styles (see Figure 2.7). The map reader can quickly identify broad areas of high and low values or compare local rates to a referent value by focusing only on one hue, but can see more specific patterns by using the full range of colors.

Brewer tested both diverging and sequential color schemes, as shown in Figure 2.7, using colors carefully chosen to avoid visual distortion due to the surrounding colors and to permit color-blind readers to use the map (Brewer, MacEachren, and Pickle 1997). Color perception is dependent on the color of neighboring areas. Different hues correspond to different wave lengths that, in turn, have different focal lengths when going through the lens of our eyes. Because red and blue have noticeably different focal lengths, shading using both colors can make the edge between shaded areas appear to shimmer. Consequently, Kosslyn (2006) recommends against shading areas using red and blue. This effect can be reduced by outlining the color in gray or black, as is done in most micromap designs. The most common type of color blindness is red-green confusion, so this pair of hues should not be used on the same graphic. However, a slight shift from these pure hues, e.g., to red-brown and blue-green, makes this pair distinguishable by all readers. The tested colors were carefully chosen so that there were equal changes of lightness between each color pair, so as not to draw attention to some colors more than others. This study demonstrated that it was possible to design color maps (or graphics) that could be used by the color blind. Brewer's website (www.colorbrewer.org) is a useful tool to help designers choose effective colors.

What the reader expects the colors to mean, i.e., color conventions, affects the accuracy of interpretation. In general, a warmer or darker color (yellow, red, or orange) is associated with higher values, and in the United States red implies danger or importance. Cultural expectations need to be considered in color choice. For example in China, red signifies good fortune and so would not be a good choice to signify danger there. In a study of map reading, readers made mistakes when these conventions were violated, such as when red represented low rates, even with legends clearly shown (Carswell et al. 1995). Similarly, the reader expects that the rank order of colors on the page is consistent with the rank order of the values they represent, i.e., the visual metaphor. An example of violation of the visual metaphor is a map legend that shows low values at the top and high values at the bottom, inconsistent with how we would read a graph.

How can we tell if colors are different enough for one group to be discriminated from others, such as for the purpose of highlighting? Bauer, Jolicoeur, and Cowan (1996) provide an answer for people with normal vision. Consider a few easily distinguished colors defined using the two-dimensional chromaticity diagram based on the CIE color system. If we wish to highlight a few items, we can encode them by a color that is not in the convex hull defined by the other colors on the graphic. This "target color" will be quickly seen using preattentive vision. An example is given in Figure 2.8—the gray dot is hard to find on the left graph, but the red dot is easy to find on the right graph.

Color has a number of merits for visual encoding, including quick perception by preattentive vision, economy of space required, and separability from other encodings, such as shape and direction. However, designers need to keep in mind the limitations and perceptual problems mentioned here in order to optimize the effectiveness of color in their visualizations.

2.4.5 Comprehension, Learning, and Working Memory

Comprehension is understanding the meaning of what we perceive. Very short-term memory, sometimes called iconic memory, stores the visual sensations, but only for up to 0.5 seconds. If it is transferred to working memory, it will remain there for up to twenty or thirty seconds, longer if we mentally repeat the information.

Working memory controls the flow of information and includes separate systems for acoustic and visually encoded information (Baddeley 1981). This dual system allows us to mentally repeat (rehearse) what we want to remember from the visualization. In simple terms visual thinking is the result of the interplay of these visual and nonvisual memory systems. We will not delve into details of nonvisual memory, which is well beyond our

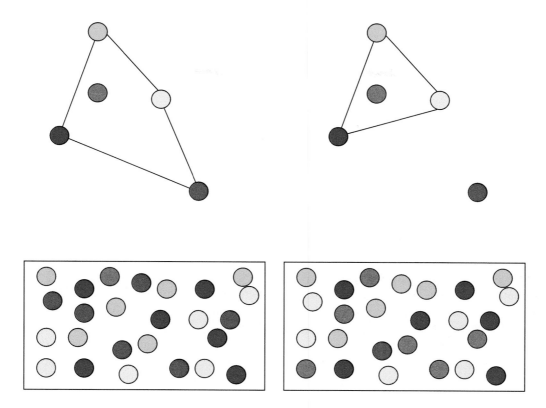

FIGURE 2.8 Example of choice of color to highlight a single item with many distracters. The diagrams are two-dimensional representations of three-dimensional CIE color space. On the left, the gray dot is hard to find because gray is within the convex hull defined by the other colors. On the right, red is outside the convex hull of the other colors and so is quickly seen by preattentive vision. (Redrawn from Ware (2004, p. 124) with permission from Elsevier.)

expertise, but note that Mayer promotes use of diagrams accompanied by spoken rather than written text, since the mind has separate visual and auditory memories (Mayer, Heiser, and Lonn 2001). Written text would compete with diagrams for visual storage in working memory, but spoken words would not.

Working memory retrieves information from long-term memory as needed to help us to understand what we placed on the "visual sketch pad" of working memory. Consolidation of information into long-term memory only occurs when active processing is done to integrate the new information with existing knowledge. What is held in working memory? Kahneman et al. coined the term *object file* to describe the temporary grouping of a collection of visual features together with other links to verbal information (Kahneman, Treisman, and Gibbs 1992). We can think of perception as occurring through a sequence of active visual queries operating through a focusing of attention to give us what we need. The neural mechanism underlying the query may be a rapid tuning of the pattern perception networks to respond best to patterns of interest (Dickinson et al. 1997). Rensink (2000, 2002) coined the term *nexus* to describe this

instantaneous grouping of information by attentional processing. Another term sometimes used to describe a kind of summary of the properties of an object or a scene is *gist*. Gist is used mainly to refer to the properties that are pulled from long-term memory as the image is recognized. The semantic meaning or gist of an object or scene (related more to verbal working memory) can be activated in about one hundred milliseconds.

In the past, paper served as an external memory to augment our limited human internal memory. Today we use computers to perform a host of data manipulations for which our tiny working memories are ill-suited. These activities include production of graphic layouts, data transformations, and encodings that seek to utilize human perceptual and cognitive strengths. What other devices can extend our working memory? If multiple data attributes are integrated into a single glyph (a visual object that displays at least one data variable), more can be held in working memory (Ware 2004). Suppose we want to represent the wind's direction, current speed, and duration at that speed on a graph or map. Remembering three separate symbols, such as an arrow with direction, color-coded circle for speed, and rectangle width

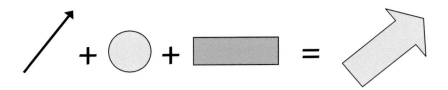

FIGURE 2.9 Combining three separable encodings—direction, color, and width—into a single glyph that is easier to remember.

for duration, requires nearly all of our working memory capacity. The same attributes can be encoded as a single glyph, e.g., an arrow with direction, where the fill color of the arrow indicates speed and the width of the arrow represents duration (Figure 2.9). These are separable encodings, i.e., each attribute can be examined separately, but the glyph is remembered as a single image so that we can remember more of them than if they were coded as three separate glyphs. The separability of an encoding method affects attention (the number of items to see), memory (the number of items to remember), and comprehension (can we disentangle the encodings to understand the full meaning?).

Another helpful method is to group three to five visual objects into perceptual groups that can more easily be saved in working memory. These limited chunks of information can more easily be processed, remembered, and compared. For visualization, we can use this to easily focus on subsets of encoded data and to scan for patterns. Cleveland (1985) suggested that dot plots be separated into logical groups, but if these are too long they should be split further. The perceptual groups are often delineated by lines or boxes, but other methods are also available, as illustrated in Figure 2.10. Perceptual grouping has been widely accepted and now commonly appears in charts in the popular press.

Grouping people, objects, and processes is natural for us. Our perceptual and cognitive systems are designed to find several different kinds of patterns. Clearly, we can enclose items to indicate their grouping. We naturally group things that are close together. Proximity grouping provides the basis for seeing patterns in the

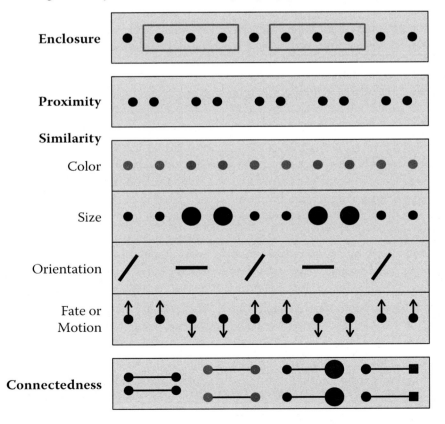

FIGURE 2.10 Encodings that enable perceptual grouping. (Adapted from Palmer 1999, p. 258, Figure 6.1.2, © Massachusetts Institute of Technology, by permission of the MIT Press, and from Ware 2004, p. 192, with permission from Elsevier/Morgan Kaufman Publishers).

dots of a scatterplot or the colors of a map, e.g., where neighboring places shaded a similar color are perceived as a cluster of similar values. Proximity can group labels with lines in a plot, as in this figure itself. We also naturally group things that appear similar, e.g., in color, size, angle, or direction. All of these features are separable encodings that can be discriminated in preattentive vision if they are sufficiently different, either in terms of type of encoding or by degree of the primary characteristic. Marr (1982) explains that we process images simultaneously at different scales. We can make easy distinctions when the stimuli are on different scales, and often the perceivable distinction has a ratio of 2:1, e.g., dot and line. Kosslyn (2006) provides more details. The last line of Figure 2.10 illustrates connectedness, perhaps a stronger grouping principle than proximity, color, size, and shape (Palmer 1999; Palmer and Rock 1994; Ware 2004). The closer vertical proximity of points in the two rows of connected pairs should also favor grouping points with similar colors, sizes, and shapes. However, the horizontal orientation and parallel structure may provide a competing bias that favors the grouping based on connectedness.

Note that we could easily remove the black outer rectangles from Figure 2.10 and still communicate which label belongs to each row of symbols. Because the separation of almost all symbols within each row is smaller than the vertical separation of the symbols between rows, and because the symbols are similar in each row, we correctly link the label with the row contents. Multiple grouping methods can operate simultaneously, such as where grouping by size also causes the larger dots to appear closer together. There are many interesting domination patterns among perceptual grouping mechanisms. For example, points within a small enclosed region will seem grouped even though they are members of two distinct larger enclosed regions (Palmer 1999).

The indentation of labels such as "Color" under the label "Similarity" in Figure 2.10 is also intended to communicate grouping, as is the removal of white space between the black rectangles in the similarity group. An alternative convention to indenting text is to center the "Similarity" label over the label space for the row below. This would make "Similarity" a title. However, such positioning breaks the alignment of "Similarity" with "Enclosure" and "Proximity," and these three are intended to be a group. A design goal for this figure was to create small perceptual groups rather than a single group with seven items. At the top level of grouping there are four groups conveying distinct concepts: enclosure, proximity, similarity, and connectedness. Then similarity shows just four subcategories, a manageable number of items for a perceptual group.

Another perceptual grouping principle, good continuation, is based on our visual ability to follow lines. That is, connecting symbols with lines not only creates a perceptual group of those symbols but also guides the eye along the desired path from one point to another. Gridlines follow this same principle, helping to guide our eye either horizontally or vertically along a graph to help us to estimate graphed values accurately, as we saw in Figure 2.5.

2.4.6 Retention, Remembering, and Long-Term Memory

Retention and remembering can be thought of as storage and retrieval of information to and from long-term memory. Retention may be faulty if we did not learn the information in the first place, or learned it incorrectly. Remembering can be improved by visual cues, such as color legends on maps or sample circle sizes on a proportional symbol graph.

Because of the capacity limitations of our memory, our brains often store information in a simplified way, leading to biases in recall. We often remember locations on a map in relative rather than absolute terms, aligning or rotating the original locations in order to better organize the map pattern to make it more memorable (Tversky 1981). Our memory of distance along a path has also been found to be biased by travel time of the route (MacEachren 1980).

Sometimes our reasoning is affected because we remember outdated or irrelevant data (Mayer, Heiser, and Lonn 2001). When asked to make an estimate, we often will start from some initial value (an "anchor") and modify this value to arrive at the final estimate (Tversky and Kahneman 1974). Once a potential customer is in the door, salespersons will typically seek to anchor their price well above a tolerable selling price; the customer negotiates from this anchor value and feels good about the final price even though it may be unreasonably high. Even totally irrelevant information can affect our estimates. When the task is to judge the number of beans in a jar, preceding that task by spinning a spinner and observing the resulting value influences people's estimates. Once we have this preconceived anchor value or image in our head, it is difficult to change it (Heuer 1999).

As noted by Ware (2004), visualization can help to extend our memory capacity. We can see thousands of encoded values in a single visual display, such as a map of cancer rates by county (Tufte 1983). We can easily remember a pattern we have noted from this image even though we could never retain many of the original numbers in our long-term memory. With computing tools, we can manipulate these views to see even more and envision the results from a host of potentially valuable perspectives.

If information of interest can be embedded within a story or something familiar, like a song, we are more likely to remember it. This idea of "stickiness" was popularized in several recent books, first by Gladwell (2000) in the best-selling *The Tipping Point* and later by Pink (2005) and Heath and Heath's (2007) book devoted to this topic. Although the term is new, most of us learned the alphabet by singing it until we got it to "stick" in our minds, and oral history has been passed down through song for eons. People are very adept at remembering stories, so if we can tell a plausible and interesting story with our graphics, they will be more memorable. We will return to this topic in the next chapter when we consider how to attract and retain the reader's attention.

2.5 SUMMARY

Visual representations can help us to think about data in several ways—by displaying a large amount of data in a small graphic, by removing some of the cognitive burden required to process the data, and by visually representing relationships in the data that would be difficult to identify otherwise. It is nearly always easier to understand patterns and relationships by seeing than by hearing or reading a description.

Micromaps achieve these goals by organizing, presenting, and allowing manipulation of data in thoughtfully structured ways. These designs evolved by combining ideas from some of the pioneers in data visualization, as described in this chapter, taking into account the constraints of human perception and cognition. Micromaps will continue to evolve as the growing interest in the analysis of data from many disciplines leads to greater cross-disciplinary collaboration and sharing of tools.

History can provide guidance for the future. In this chapter we have highlighted advances in statistical graphics, cartography, cognitive psychology, and computer science and examined our cognitive strengths and weaknesses that are relevant to visualization design. In the next chapter we apply this knowledge to derive design principles for data visualization.

FURTHER READING

THEORETICAL BACKGROUND

Cleveland, W. S. 1993. *Visualizing data.* Summit, NJ: Hobart Press.

Dent, B. D. 1993. *Cartography: Thematic map design.* 3rd ed. Dubuque, IA: Wm. C. Brown Publishers.

Herrmann, D. J., Yoder, C. Y., Gruneberg, M., and Payne, D. G. 2006. *Applied cognitive psychology.* Mahwah, NJ: Lawrence Erlbaum Associates.

Palmer, S. E. 1999. *Vision science: Photons to phenomenology.* Cambridge, MA: MIT Press.

Ware, C. 2008. *Visual thinking for design.* Burlington, MA: Morgan Kaufmann Publishers.

Wilkinson, L. 2005. *The grammar of graphics.* 2nd ed. New York: Springer.

APPLICATIONS OF VISUALIZATION THEORY

Brewer, C. A. 2005. *Designing better maps: A guide for GIS users.* Redlands, CA: ESRI Press.

Few, S. C. 2009. *Now you see it.* Oakland, CA: Analytics Press.

Heuer, R. J., Jr. 1999. *Psychology of intelligence analysis.* Washington, DC: Central Intelligence Agency. Available from https://www.cia.gov/library/center-for-the-study-of-intelligence/csi-publications/books-and-monographs/psychology-of-intelligence-analysis/PsychofIntelNew.pdf.

Kosslyn, S. M. 2006. *Graph design for the eye and mind.* New York: Oxford University Press. (Note: We often reference Kosslyn's 1994 edition of this book because that was what Dan used in his early research. However, we point the reader to this new edition for further reading.)

MacEachren, A. M. 1994. *Some truth with maps: A primer on symbolization & design.* Washington, DC: Association of American Geographers.

Monmonier, M. 1993. *Mapping it out: Expository cartography for the humanities and social sciences.* Chicago: University of Chicago Press.

Robbins, N. B. 2005. *Creating more effective graphs.* New York: John Wiley & Sons.

Stafford, T., and Webb, M. 2004. *Mind hacks: Tips and tricks for using your brain.* Sebastopol CA: O'Reilly.

Tufte, E. R. 1983. *The visual display of quantitative information.* Cheshire, CT: Graphics Press.

Tufte, E. R. 1990. *Envisioning information.* Cheshire, CT: Graphics Press.

3 Data Visualization Design Principles

Solving a problem simply means representing it so that the solution is obvious.

—Herbert Simon (Simon 1996; Thomas 2007)

3.1 INTRODUCTION

Remembering an example where a graphic representation clarified a situation or resolved a problem, many of us might be inclined to agree with this statement. However, finding a visual representation of data that perfectly resolves the question at hand is no simple matter. The underlying quantitative relationships are rarely simple, and there is a tension between design goals, such as providing context for meaningful interpretation and simple appearance. As we bring more and more content into a graphic it appears more and more complex, to the point where the graphic becomes cluttered and confusing to the reader. Sometimes the complexity of supplementary information contrasts sharply with the simplicity and sparseness of the data, as in a recent example of a cancer statistics plot, where seven lines of footnotes described important details about six data points for oral cancer among females 2000–2005. Ideally, adding features to the graphic can make the pattern seem simpler or provide a focus to prioritize the information, but we need to be careful not to overly complicate the graphic ("chart junk"; Tufte 1983) (National Cancer Institute 2008a).

Good design seeks to address challenges posed by characteristics of the data, the tasks to be performed, and the skills and limitations of the reader. These challenges exist in varying degrees for all types of graphics, from simple printed plots to dynamic graphics views. Data-related challenges include visually representing massive data sets of multivariate observations that may vary over spatial and temporal scales and attribute indices. Data limitations of content and reliability also constrain our graphical displays to what is available, not what is most desirable. For example, we would like to have cigarette smoking rates (ideally categories of prevalence by duration and amount) by county and gender starting in the 1950s to correlate with lung cancer rates decades later, but these data do not exist. Surrogates such as cigarette sales data are poor correlates of smoking behavior. We cannot wave our magic wand and make

the ideal data available, reliable, or representative of the population. We usually just have to make do with what we have. The questions asked of the data—reading values, identifying patterns, or comparing patterns— also impact the design. Finally, as we learned in the previous chapter, humans have powerful preattentive and unconscious processing abilities, but we exhibit different kinds of blindness, such as inattentional and change blindness, we have tiny working memories with correspondingly limited calculation abilities, and we are prone to well-documented reasoning flaws, such as anchoring to an incorrect value (Tversky and Kahneman 1974) or remembering graph curves as more symmetric than they really were (Tversky and Schiano 1989).

The graphics designer must find a compromise between simplicity of design that accurately conveys the underlying data patterns while acknowledging human perceptual and cognitive limitations and the complexity of real-world data. In this chapter we apply the research summarized in Chapter 2 to propose effective visualization designs for the task at hand.

3.2 ENABLING ACCURATE COMPARISONS

The goal of enabling accurate comparisons motivates our selection of design features used in linked micromaps. Visual representations that are easily understood will stimulate mental comparisons against what we know as we endeavor to make sense of what we see. Well-designed graphics allow us to make rapid visual comparisons that reveal patterns. Returning to the baseball example from Chapter 1, we could explain to you in words how well players at each position hit and fielded, on average, in 2007. In order to think about spatial patterns in the data, you would have to first recall where each player is normally positioned on the field, then compare the performance statistics between players on the left or right, between players in the infield or outfield, and so on. This could easily take a few minutes of concentration. One glance at the linked micromap plot (Figure 1.5), though, shows that players on the left and right sides of the field were better batters than those in the middle. Even people unfamiliar with baseball positions can see this pattern.

In this section we discuss how to enable accurate comparisons by our graphic design and style choices. For the purpose of discussion, we define design choices as those options based on task prioritization and guidance from cognitive research and style choices as preferences of lesser importance that may not be well studied in the literature. A secondary purpose of this section is to encourage continuing thought and research on task-based comparative graphics.

3.2.1 Comparisons Using Position along a Scale

We start our examination with a look at comparison of attributes, i.e., characteristics of our observations. Attributes are important! Without attributes we would know nothing about the places on our map or about the people who live there. Processes play out in space and time, which are powerful descriptors but are not usually regarded as agents of change. We are willing to sacrifice some accuracy in representing space and time but, because of their importance, we want to represent the attributes accurately. In this book we typically use the term *variable* to refer to a vector of attribute values, with each element of a vector corresponding to a particular study unit, such as a county. Variables can represent either the outcome (study variable) or potential predictors of it. In multiple linear regression, the dimension of the linear vector space for the predictor variables is determined by how many of the predictor variables are linearly independent. While there are practical limits, such as storage capacity and the increasing number of observations needed to support independent vectors, there is no conceptual limitation on the dimensionality of the vector space. Some of today's data sets contain millions of variables that could be used as predictor variables in a regression model. Of course, we should show only the most relevant variables on the display, but we often need to allow the analyst to choose a reasonable number of variables from a much longer list. We assert that attributes and their values should not be considered as second-class citizens relative to space and time.

The most accurate direct visual comparisons are those made along a common scale (Cleveland 1985). We will use simple bar plots of a continuous attribute variable to illustrate the basic design issues for these fundamental tasks. One historical issue for bars has been the choice between vertical and horizontal bar orientation. Some early cognitive research gave the edge to vertical bars, but more recent research indicates very

little difference in terms of decoding accuracy (Kosslyn 1994a). Our choice is almost always horizontal bars with labels on the left side of the graph because reading left to right will lead our eyes to the plot. We are not alone in our reasoning that in English it is easier to read the horizontal bar labels, rather than vertical or slanted labels (Few 2004). Because people in English-reading cultures have a tremendous amount of practice reading left to right, there are performance benefits in addition to benefits of familiarity and expectation. For cultures that read vertically, we suspect vertical bars would be preferable. For vertical labels that are normally read from top to bottom, putting the labels above the bars would promote reading into the bars in the plot rather than away from the plot.

A second issue concerns the alignment of bar labels. Should labels be left aligned as expected in conventional text, centered as might occur in poetry and appeal to our sense of symmetry, or right aligned to be closer to the bars, exploiting proximity-based perceptual grouping with the bars? Our guess is that this is a close call. We lean toward left alignment because it seems to facilitate seeing groups when there is a purposeful vertical gap between groups and because we sometimes scan the labels vertically. We suspect that left alignment is helpful for reading in this context. There is a convention to align text on the left, integers on the right, and floating point numbers by the decimal place. We tend to follow convention because of reader familiarity unless there is a strong cognitive reason to break convention.

A third issue is whether to outline the bars. Few (2004) recommends not outlining bars if they already have adequate contrast with the background. If an outline is needed, for example, to separate yellow from white, he suggests using a gray outline rather than black to avoid extreme contrast. We think this is a close call and frequently have used a black outline to bring bars and maps into the foreground.

As we saw in Chapter 2, grid lines can help the reader to accurately extract values. The bars should be in the foreground and grid lines in the background, but should the grid lines be visible through the bars? Options include translucent bar fill or totally covering the grid by the bars. A compromise may be the best we can do. We don't have a strong recommendation other than making sure that the end of the bar is clearly seen.

Few (2004) provides guidance on several other facets of bar plot appearance. He prefers roughly equal widths for bars and the space between bars when the bars belong to the same group. Some bar encodings complicate decoding, such as using the width and length of a rectangle to encode two variables. Cognitive scientists have found that people primarily respond to the area of the rectangles and have a very difficult time decoding width and length separately. Such encodings are called integral encodings because people respond to the composite and have to work hard to get at the parts for purpose of separate comparison. Color is another integral encoding. Our eye-brain system is not designed to decode three variables encoded using the color dimensions such as hue, lightness, and saturation.

Some encodings are separable. That is, as we discussed in Chapter 2, each attribute can be interpreted separately, but we can more easily remember the composite image, like the color-coded arrow discussed in Section 2.4.5. For example, our perceptional systems can separate the color of a line from the length of a line. If there are many red and blue lines, we can focus immediately on the red lines and compare lengths of red lines. We don't have to stare at individual lines and think consciously about separating color from length. As indicated above, color is a poor encoding for a continuous variable, but a few discriminable hues work well encoding a categorical variable with just a few categories. Some clever designers may find a way to encode three variables using color, but it usually is neither faster than nor as accurate as separate encodings. A ray with a dot at the base has two encodings—length and angle—that are both good encodings individually. The situation is not so clear about how separable they are, since the ray length must be long enough to detect the angle. Kosslyn (1994a) indicates that angle differences of 30° are easily discriminated. A few well-separated angles could also be used to encode a few categories.

3.2.2 IMPROVEMENTS TO THESE BASIC DESIGNS

As we saw in Figure 2.5, plots are usually improved by the addition of background shading with grid lines to help the eye accurately judge the values on the graph. Sorting the bars can also help to quickly judge rank order or make comparisons among the bars.

Somewhere between design requirements and style preferences there is a domain of choices that make some difference but often not a big difference. One of these is the treatment of tick marks. Cleveland advocates putting the tick marks outside the panel to prevent plotting symbols on the tick marks. As the middle panel in Figure 3.1 suggests, we can drop tick marks when we use grid lines. The printed page and display screen have only so much space, and so graphics designers are engaged in a continual battle to obtain visual real estate. Our opinion is that the middle panel appears simpler. Our visual system processes edges well, and so we conjecture that the ticks draw attention beyond their value. Conventional software that puts tick marks inside the panel still puts the tick mark labels outside the panel. We are convinced that alignment of the grid lines with the grid line labels and their proximity can induce the proper perceptual grouping of labels to grid lines without the need for tick marks.

We are also interested in providing a sense of value added. Might overplotting tiny white dots help locate the center of the dark dots and lighten the image? The right panel of Figure 3.1 provides an example. Our impression is that the doubly plotted dots appear slightly smaller as a whole, that we are more confident about the precise location of their centers, and that the dots appear lifted a little off the page (i.e., more in the foreground). This little three-dimensional cue moves the dots closer to

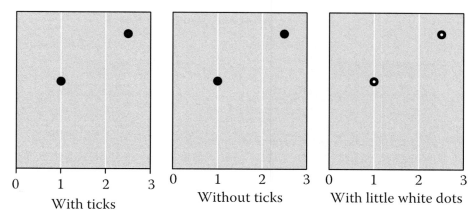

FIGURE 3.1 Here the middle panel saves space by removing tick marks and the right panel adds little white dots as an enhancement.

us and further distinguishes the dots from the gray fill and grid lines in the background. Adding the dots may just be a style choice, but conveying the feeling of confidence is good. Further, the small three-dimensional effect can give the impression of value added that is also good as long as it does little harm. We will occasionally add little three-dimensional cues like this.

3.2.3   Zooming in to the Data

Sorting the rows helps us to mentally group bars with similar values together and to make comparisons among them. In Figure 3.2, the lung cancer rates for selected states are sorted by white male rates. Several patterns stand out. For example, the longest bar for men (Louisiana) is much longer than the shortest bar (Utah).

The width of the panel for white women is substantially smaller than the width of the white male panel. As we add columns to the page for additional variables, horizontal space becomes precious. If we used the same scales and panel width for both the white male and white female panels, the right side of the white female panel would be empty. We can crop the panel on the right so that it is just wide enough to enclose the bars and then reallocate the saved space to both panels. Technically, the panel construction set uses the scale values separately for men and women and then allocates panel widths proportional to the maximum scale values. Consequently, the data units (mortality rates) per unit length (inches or centimeters) are the same for both panels, making the bar lengths directly comparable. This is one way to zoom in on the data—cropping the panel to be just large enough for the bars.

Since we want to encourage comparisons between the male and female rates, the bars are ordered in the same way for both, even though this means that the female bars are not in increasing rank order. If we disconnected the two panels, labeling the male and female bars separately, we could sort each panel to rank order the bars, but then in order to compare the male and female rates the reader would need to scan each panel to find the pair of bars for each state. This would increase the cognitive burden on the reader, so for most tasks the row-labeled plot design (Figure 3.2) is preferred.

3.2.3.1   Scales and Scale Labels

In this book the task is to show data-based variation, and so graphics resolution is important. (We will discuss resolution a little more when we address slider construction.) Graphics best serve this task when we choose suitable scales. In *The Grammar of Graphics*, a brilliant conceptualization that supports the algebraic construction of graphics, Lee Wilkinson (2005) devotes an entire chapter to scales. With regard to whether the axis should always include zero, he writes, "Money is not a physical or fundamental quantity, however. It is a measure of utility in the exchange of goods. Research by Kahneman and Tversky (1979) has shown that zero (no loss, no gain) is not an absolute anchor for monetary measurement" (p. 91). Wilkinson seeks a reasonable ground in terms of selecting scales, writing, "Since we are not God, the best we can do is understand the context in which the data were collected and use our intuition and substantive scientific knowledge to choose scales" (p. 89). We agree.

Wilkinson also discusses criteria and an algorithm for choosing nice axis scale values for ticks or gridlines.

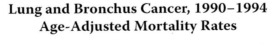

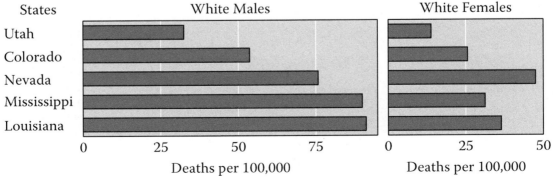

FIGURE 3.2   Bar plots for comparing lung cancer mortality rates for white males and females in 1990–1994 for five selected U.S. states, displayed in separate panels to conserve space.

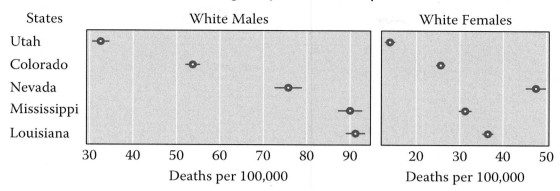

FIGURE 3.3   More detail can be seen by using dot plots (shown with 95% confidence intervals) than the bar plots in Figure 3.2 because the scale covers only the data range.

Some examples in this book could be improved by a better algorithm that would search more diligently for nice scales. He strongly prefers axis scale values whose first and last values are at the end of the scale and hence align with the edges of a plot or panel, citing the cognitive literature supporting his position. In this book we require that the scale limits be big enough so dots and other plotted symbols do not plot on or outside the panel frame. We often juxtapose panels (plots), so we want grid line labels to be far enough from the edges of panels to have a gap between them. In graphics design there are often ideals that conflict and choices to be made from alternative compromises. Instead of using internal grid line labels we could leave a gap between panels, plot scale values at different distances from the panel edges for every other panel in a sequence, or plot values on opposite sides of the panel for every other panel. These solutions take up precious space and the last two complicate plot appearance.

3.2.3.2   Dot Plots

Unless it is important to include the zero value in the context of our data, we can zoom in to the data range in each panel and use the individual scale limits to set widths for comparable units per inch across the panels. Because classic bar plots imply a length encoding, they need to start at zero. However, focusing on just the range of the data often gives much better resolution. This leads us to use the row-labeled dot plot design illustrated in Chapter 1. This design is common for plots of stock market data, for example, which typically show the daily range (high and low) as well as the open and close prices for a stock. Very few of the designs in this

book will be bar plots, since we prefer scales for our plots based on the range of the data.

Figure 3.3 shows lung cancer mortality rates for the same data as in Figure 3.2. Reading the x axes makes the range differences clear. This design saves space and increases resolution. Sometimes we may choose to use a common range for all of the panels, approximately ten to ninety-five for these data. This can be helpful for making comparisons, but the trade-off is a less efficient use of space.

Dots representing mortality rates in Figure 3.3 are plotted on top of 95% confidence intervals. The confidence intervals are small relative to the ranges of the rates. The rates are well separated with no overlap of confidence intervals except for the male rates of Louisiana and Mississippi. The nonoverlapping confidence intervals tell us that the ranking of state values is well established and not likely due to chance. The high rate for women in Nevada stands out as an anomaly, as it did in Figure 3.2.

3.2.4   Time Series Data

Showing temporal patterns provides another basis for comparison. Arrows work quite well for showing comparisons involving pairs of values with the same units of measure. Tversky (2005) has shown that most people use arrows to explain directions and so instantly understand their use to depict changes in space and time when used in a graphic. Here, an arrow's two ends show values for the two time points, providing context using position along a common scale. The arrow line connecting the endpoints encodes the magnitude of the change

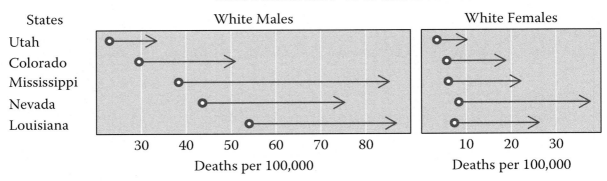

FIGURE 3.4   Arrow plots show change over time.

using length, a good encoding. The arrow tip shows the direction of the change.

Figure 3.4 shows the change in lung cancer rates from 1950–1969 (dots at end of arrows) to 1970–1994 (arrowheads). The panel widths match the range of values for males and females so that the units per inch are comparable. The lengths of arrows can be compared both within and across panels.

The wider panel for white males catches the eye—this is needed to capture the large increases in lung cancer rates over the period. For white males, Mississippi has the largest increase and nearly equals the Louisiana rate by the second period. For white women, the Nevada rate shows the largest increase, with a slightly higher rate than other states at the beginning but much higher by the end.

This example illustrates the efficiency of the arrow plot, which shows the two time-specific values and differences on a single graphic. We could calculate the differences and display them directly, e.g., a 50% increase in a state's rate. This direct difference plot has only half as many points plotted, enabling observation of finer details, but it removes the context provided by the values of the original pairs. For example, if the Utah rate doubled over a decade, it would stand out on a difference plot but we would not see that its rates are still among the lowest of these states.

How can we display multiple time points? A natural extension would be to replace the arrow with a small time series graph line for each row. This is an effective display if the reader is interested mainly in the general time trend, but can be too complex or too small to show much detail (i.e., comparing curves is much more difficult than comparing points or arrow lines). If the primary interest is in the spatial patterns, then animated maps might seem to be ideal, but as we learned in Chapter 2,

our attention and memory are too limited for this to be effective for any but the simplest maps. An exception is a map of a spreading process, such as of a contagious disease from a single source. Then we can focus on (attend to) the initial point source and see how the pattern expands from that point. However, if there are multiple points of interest or places with rapidly changing values, then we are unable to see all of the pattern changes at once. A third technique for displaying time series data is to present time-specific maps simultaneously, usually all on the same page (small multiples), so that the reader does not need to remember the previous map pattern, as is necessary for animation. This method has been demonstrated to lead to more accurate data extraction than animation (Robertson et al. 2008). A further enhancement is to supplement these maps with maps of the calculated differences in values between adjacent time points. This reduces the cognitive burden on the reader by displaying the original values plus the changes on one graphic, much as the arrow plot does for two time points per geographic area. This concept is the basis for comparative micromaps.

3.2.5   MAP PATTERNS AND COMPARISONS

We want to design our graphic so that extracting values from it is most accurate, and we have seen that this is best done by reading values from an aligned scale. When data are mapped, the geographic units cannot be manipulated by alignment or sorting as we have done with bars or dots on a plot. What design elements of a map will lead to the most accurate pattern recognition and comparisons?

Most of the cognitive experiments on map reading conducted at the National Center for Health Statistics, summarized in Chapter 2 (Sections 2.3 and 2.4.4), used accuracy of rate readout, pattern recognition, or pattern comparison to rank the options for each map element design. The most accurate design was a choropleth (area shaded) or a smoothed map style, depending on the task, with a standard vertical legend, e.g., Figure 2.3. All values were categorized by quantiles (e.g., quintiles) and represented by a sequential or diverging color scheme that even the color blind could use. These experiments were conducted on mortality rate data for the purpose of designing paper maps, so further experiments on web-based, interactive maps could modify these recommendations. However, these recommendations have been used for several web applications and seemed to perform well. They provide a reasonable starting point for map design.

Beyond these basic recommendations, we have found that analysts need maps constructed using familiar geographic units or clear help needs to be provided for geographic context. County- or state-level maps have easily recognized boundaries, but even when these common geographic units are combined, such as to achieve a minimum population size, the new units may no longer be identifiable. As an example, health service areas (HSAs) used for the NCHS atlas were constructed by aggregating counties based on where Medicare patients received hospital care (Makuc et al. 1991). The atlas included a map with HSA numbers that were linked to a list of the largest city and counties in that area, but still readers found it difficult to flip back and forth to the referent map to match landmarks to the mortality map. There was nothing to make the definitions stick mentally. The boundaries may be smoothed (generalized) for presentation as long as the geographic units are still recognizable.

The eye is drawn to large blocks of bright (highly saturated) colors. Since large geographic areas typically are sparsely populated, our visual impression of the map pattern can be biased by the color choices. Cartograms, maps where the geographic units have been distorted so that area is proportional to some characteristic, often population, may remove the large area bias from the original map, but they cause confusion among analysts who can no longer identify landmarks, cities, or even adjacencies on the distorted map. We prefer to use boundaries that are recognizable and to use colors that are carefully selected so that none dominate the map (see www.colorbrewer.org for color selection tools), unless we wish to draw attention to only a few areas.

For a map covering a large geographic area, as for all U.S. states, an appropriate map projection must be chosen. Most cartographers prefer a projection that preserves areas, such as Albers equal area projection, in order to avoid areal bias, but there may be circumstances when other types of projections would be appropriate. The U.S. state map design used for micromap plots in this book is a simplification for better visualization. We discuss this choice further in the smoothing section (Section 3.3.3). Note that some U.S. states have their own projection, usually one that will display the state and its counties in a more familiar orientation than when mapped at the U.S. level.

The need to aggregate data to display rates or other summary statistics on a map leads to what is called the modifiable area unit problem (MAUP) (Openshaw and Taylor 1979). Aggregation to one set of administrative boundaries versus another can lead to different statistical summaries and apparently different geographic patterns on the map (Gotway and Young 2002). Further aggregation of small geographic units (small in population or area) to larger ones in order to create more reliable estimates or more visible map units can also cause different map patterns. Box plots of the nested units, e.g., counties within states, will show how variable the smaller units are, thus pointing to larger units that may not be represented well by a single summary measure, such as the mean.

Because of this problem, the analyst should compare maps at different scales, if at all possible, before drawing conclusions about patterns. To illustrate, Oliver et al. (2005) found very different geographic patterns of prostate cancer incidence in Virginia when the rates were mapped by census tract versus county. Further investigation found that geocoding error rates were much greater at the tract level and varied by urban-rural status, biasing the patterns on the tract maps.

The choice of most appropriate geographic boundaries is further complicated for a series of maps, such as for a time series. Administrative boundaries occasionally change, so it is important that the map boundaries are consistent over the entire range of maps. For example, La Paz County, Arizona, was created by splitting Yuma County on January 1, 1983. A time series of maps with data prior to 1983 could not show La Paz and Yuma counties separately in order to be consistent over the entire time range.

3.2.6 HYPOTHESIS GENERATION

What explains the variation in lung cancer mortality rates for the five selected states in Figure 3.3? Is it just random variation? Probably not, since the populations are large and over 50,000 deaths occurred in these states, as we can see from the narrow confidence intervals. Epidemiologists know that cigarette smoking is a major cause of lung cancer. Does the variation in cigarette smoking rates explain most of the variation in the plot, or are there environmental or other behavioral factors that might impact lung cancer mortality independently or through an association with cigarette smoking? Patterns often lead to questions. We will illustrate this process of hypothesis generation using these lung cancer data.

The pattern in the white male panel for 1970–1994 (tip of arrows in Figure 3.4) indicates a substantial difference in mortality rates between Utah and Louisiana—about fifty-two deaths per one hundred thousand. As a benchmark for comparison, this difference is greater than the U.S. mortality rate for every type of cancer except lung cancer. The order of the Nevada, Louisiana, and Mississippi rates is reversed for white females compared to the white male ordering, due to the relatively higher female rate in Nevada.

What factors might be involved? We know that the primary risk factor for lung cancer is smoking, both active and passive, with elevated risks also found for certain occupational exposures, such as to asbestos and arsenic, and general environmental exposures, including air pollution, ionizing radiation, and radon in the home (Alberg, Ford, and Samet 2007). We also know that lung cancer typically takes at least twenty years to develop, so any hypothesized cancer-causing exposures must have occurred decades before. Unfortunately, smoking rates for all U.S. states and counties are only available from surveys conducted in the recent past (Centers for Disease Control and Prevention 2003; Bureau of Labor Statistics, Census Bureau, and National Cancer Institute 2009). Information on occupational exposures is also limited to survey data aggregated to the state or county level—we do not have routinely collected information on what occupations each resident has had, nor do we have residential histories for everyone that would place their exposures in a spatial context.

Given these data limitations, what do we know about the states involved? Utah, Nevada, and Colorado are neighboring states in the West with similar weather and environmental conditions but different levels of smoking. Utah has a large Mormon population and their religion opposes cigarette smoking; in 2007, 30% of Utah residents claimed they had ever smoked and only 12% were current smokers (Centers for Disease Control and Prevention 2008). Nevada is known for its

gambling industry, where exposure to passive smoking is high in casinos, which are exempted from a statewide smoking ban (Benston 2008). In 2007, 47% of Nevada residents said they had ever smoked and 22% currently smoked. Colorado is less extreme in terms of Mormon religion- and gambling-related influences and has intermediate levels of current smoking (43% ever smoked and 19% currently smoke). Mississippi and Louisiana are neighboring states in the South that had more families living in poverty than any other states in 2000, and much lower than average education levels. Both low education and low income are correlated with cigarette smoking—the percent of ever smokers in Mississippi and Louisiana is similar to the U.S. average (45% and 43%, respectively, compared to 45% for the United States), but more residents currently smoke (24% and 23%, respectively, compared to 20% for United States). News reports related to flooding after Hurricane Katrina called attention to the gambling industry along the Gulf Coast, and so perhaps passive smoking exposure plays a role in these states. Other factors that these Gulf Coast states have in common include their weather—could high humidity and hot weather be associated with breathing difficulties, lower activity levels, and increased smoking rates? Just a little knowledge about the states makes it easy to generate ideas about variables, but we need to be careful not to confuse associations with causality. Our hypothesized risk factors could be directly, indirectly, or just coincidently associated with lung cancer mortality.

3.2.7 Transformations

Many comparative graphics exploit the power of data transformations. As psychologists have shown, most of us are not good at transforming data in our minds and envisioning the results. If the difference between paired values brings out important patterns in the data, it is important to represent the differences directly and not leave the task to the analyst. This is a good example of what it means to reduce the reader's cognitive burden and is a technique we use for comparative micromaps. Even with the best design, we have trouble calculating mental differences.

Curve comparison is a common quantitative task that almost always involves assessing difference in the vertical direction. However, our eyes evolved long before visual quantitative analysis, and so we tend to assess the closest difference between curves (see left panel of Figure 3.5). Adding vertical grid lines in the middle panel helps us to see the vertical differences. A few grid

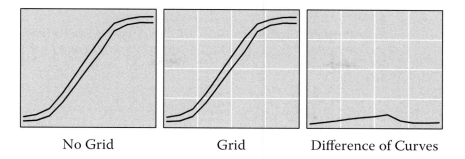

No Grid   Grid   Difference of Curves

FIGURE 3.5   Grid lines help in accurately seeing the difference between curves.

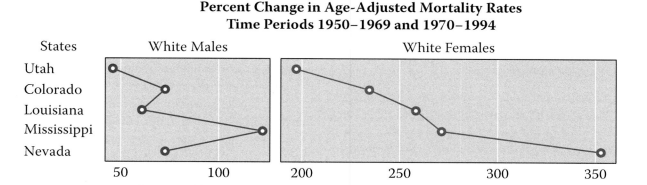

Lung and Bronchus Cancer
Percent Change in Age-Adjusted Mortality Rates
Time Periods 1950–1969 and 1970–1994

FIGURE 3.6   Connecting the dots vertically draws attention to anomalies in percent change in lung cancer mortality rates from 1950–1969 to 1970–1994.

lines help but may be too sparse to adequately reveal the pattern in the difference of curves. The right panel shows the difference between curves. The size of the maximum difference can be a surprise. Imagine how much more difficult it is to mentally calculate differences between maps than in this simple graph example.

We can transform values in ways that provide a different way of thinking about the data. A common transformation is percent change. Figure 3.6 shows the percent change in lung cancer rates from the aggregated time periods 1950–1969 to 1970–1994. Our attention is drawn to the 350% change for white females in Nevada and to Mississippi for white males.

As previously noted, the percent change transformation halves the number of values presented. We can compute the ratio of female to male percent change to further reduce the number of values to 5, one per state. For example, the ratio in Utah is roughly 200%/50% = 4. While the transformations lead to fewer numbers to visualize, each transformation takes us further from the original data, i.e., the numbers of people dying from lung cancer. The ratio of rates is unitless and so is easy to forget. Adding a second panel with the number of deaths would provide an important reminder of the number of preventable early deaths.

In some circumstances taking logarithms of the values is helpful, such as when the distribution of positive values is very skewed. A log transform will pull in the long tail of the distribution, making the values and rank orders easier to read from the graphic. On the other hand, if your audience has low quantitative skills, displaying logarithms can be off-putting.

The lung cancer mortality data plotted here have already been transformed from case counts to age-adjusted rates. This transformation endeavors to make states comparable by removing the strong dependence of cancer rates on age so that we can see patterns due to other, more interesting factors. Without age adjustment, for example, crude death rates for Florida will be much higher than those in Utah because of the many older retirees in Florida.

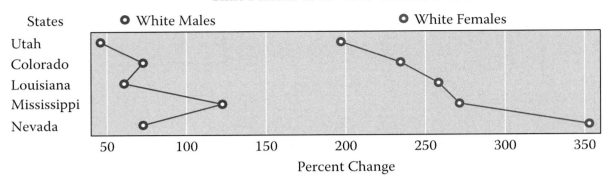

FIGURE 3.7   Figure 3.6 percent change in lung cancer mortality rates from 1950–1969 to 1970–1994, redrawn in a single plot frame (superposed panels).

3.2.8   PANEL ARRANGEMENT

So far, we have confined our comments mostly to the design of a single panel, such as the white male lung cancer rate dot plot in Figure 3.3. Once we have devised a good design for one panel, we need to decide how to arrange multiple panels on the page or screen.

We start by following Cleveland (1985) in calling attention to two comparative panel layouts: *juxtaposed panels* and superposed panels. Juxtaposed panels are side-by-side panels typically distinguished by values of one or more categorical variables such as race and sex (e.g., Figure 3.4). *Superposed panels* have subsets of data indexed by categories in a single panel, e.g., male and female cancer rates on the same graph, that are distinguished by different colors, shapes, and additional labeling. Superposing panels can also involve adapting individual panel sizes and scales for compatibility in a joint panel. To illustrate, we have combined the male and female panels of Figure 3.6 into a single panel in Figure 3.7.

Several design criteria can influence the choice between juxtaposed and superposed panels. Rich graphics often enable more than one kind of comparison. One important design criterion is to facilitate the comparison task of greatest importance. Two more important criteria are controlling visual and logical complexity, which we discuss in the next section. Further considerations can include supporting comparisons from multiple perspectives and making good compromises in light of the media constraints and audience characteristics. As we can see from Figure 3.7, superposing the rate graphs makes it easier to see differences between the male and female rates. On the other hand, the need for a

single-rate axis that spans the range for both males and females reduces the resolution of the graph, compared to the separate panels in Figure 3.6, where we focused each panel axis on the range of only that panel's data.

Panels in a composite graphic may be arranged in several ways. Aligning separate panels vertically can facilitate comparison of values on a common scale by reading down the panels, such as when perceptual groups are arranged one above the other. Aligning the panels horizontally facilitates reading values across the panels, as in row-labeled plots, where each panel corresponds to a different attribute but with the rows having a common label on the left of the panels. Horizontal and vertical alignment can be combined to create a two-way layout of panels, much as one creates a two-way table layout. One familiar example of a two-way graphical layout is the scatterplot matrix. We will discuss the two-way layout further in Chapter 5, where it is used to partition a choropleth map by two attribute variables.

3.3   STRIVE FOR SIMPLE APPEARANCE

[Simplicity] comes from the intelligent desire for clarity that gets to the essence of the issue…. Simplicity is not easy, in fact, it is hard.

—Garr Reynolds (2008)

Everything should be made as simple as possible, but not one bit simpler.

—Albert Einstein (2009)

The first quotation connects simplicity with issues. We want to develop an understanding of a complex world based on quantitative relationships using our limited mental capabilities. Some of today's data sets involve terabytes of data, and our working memories can only deal with four chunks of information or less at one time. Quantitative design issues swirl around bridging the huge gaps between data and human goals, models, tools, and capabilities—not a simple task. When we choose to show complex relationships, all we can do is strive for simple appearance as we endeavor to help people (including ourselves) understand the data. Simplicity of a graphic's design or the story it conveys will also help the reader to remember it (Heath and Heath 2007).

In the face of complex quantitative relationships, simplicity is about starting from where people are and taking modest steps toward complexity. What is simple for some people may be difficult for others because they are starting from different places in terms of culture, background, training, ability, and interests. There is a lot of wisdom in the common guidance that says "know the audience." Nonetheless, our design guidance should be given along with an admission of humility about how much we don't know about the audience. Striving for simplicity in the face of background noise, overwhelming detail, and complexity is hard work. In this section, we examine the concepts of perceptual grouping, aggregation, and smoothing that can help us to simplify our visualizations.

3.3.1 PERCEPTUAL GROUPING AND ALIGNMENT

The principles of perceptual grouping summarized in Section 2.4.5 can be used to simplify the appearance of a micromap and thus lead to more accurate use. In our plots we primarily use enclosure, color, proximity, and occasionally connectivity to create perceptual groups. Enclosing each micromap panel and group of panels clearly delineates the items in each group. Color is used to group items across each row of a linked micromap, but the colors on rows within each panel are very distinctive so that the reader will not think that the regions within the panel are similar (except by sorted values). In conditioned and comparative micromaps, people with normal color vision can focus rapidly on a single group of regions if their colors are similar to each other but different from other nearby regions. Performance (speed and accuracy) drops as the number of colors increases (because the colors will necessarily be more similar) and as the number of distracters increases, as we saw in the "eureka" example in Chapter 2 (Figure 2.4).

How many items or chunks of information should be in a perceptual group to keep a simple appearance? Kosslyn (1994a) cites much research that indicates the magic number is four. If this is correct, how can we support the visual discovery and communication of quantitative patterns for potentially very large data sets with many terabytes of data? With increasingly large data sets and complex models, e.g., with hierarchical structure and spatial autocorrelation, we become increasingly dependent upon computers and models to find and summarize patterns. Our goal is to push back the bounds on effective quantitative visualization by following the guidance from researchers summarized in Chapter 2 so that we can see patterns in the data at a high level even if we can't cope with the mass of details. Perceptual grouping is one way to achieve this goal.

Perceptual grouping by proximity is used in micromap design to link labels with their graphical counterpart, such as for the rows of linked micromaps. If we think about our common experience reading text, the white space between lines helps us to be confident about what is a single row. We can use a smaller white space between characters in words than between lines of words to provide a proximity basis for joining characters horizontally. However, even if there are larger gaps between words in a line compared to gaps between lines, we tend to read along left to right, grouping the words horizontally. People whose native language is read vertically would most likely tend to group vertically. This tendency to continue reading horizontally or vertically may be due to perceptual grouping by proximity or by continuation, where our eyes are led in a certain direction by some connectors (see Section 2.4.5). Whatever the rationale, we will make heavy use of horizontal and vertical alignment in our graphic designs.

Figures 3.6 and 3.7 use good continuation to emphasize the state-to-state differences reading down the panels. Even though an analyst might expect that the percent changes in lung cancer rates will be similar for males and females, after sorting states by the female percent differences, the vertical lines help us to quickly see that the male percents are not ranked the same. Our eyes are drawn to the long lines and opposite directions for the lines for Mississippi and Nevada as we read from top to bottom on the male panel. The correlation between percent changes for males and females could augment the plot. Most of us are not adept at mental calculations, but if graphics are well constructed, we can see some basic patterns and anomalies.

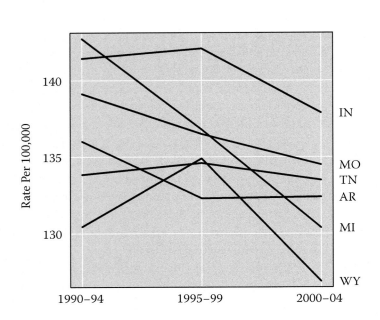

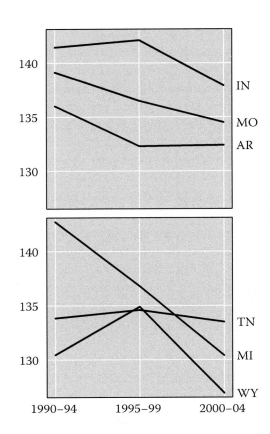

FIGURE 3.8 Simplifying appearance by perceptual grouping. Comparisons are much easier using the graphs on the right than the very busy graph on the left. Data are all cancer mortality rates among white females by time period and state.

Kosslyn (1994a) provided a dramatic example of simplification by perceptual grouping. We have redone his example in Figure 3.8 by making the total area of the two rightmost graphs equal to the area of the graph on the left. The plot on the left looks very busy, but why? One measure of complexity is the number of comparisons we could make using the plot. Here we have six lines, so there are potentially fifteen different paired comparisons (six choose two), well beyond the three or four things we can easily keep in mind at one time. Each of the right panels has only three lines, so there are only three comparisons in each, a much easier task. These panels look simple.

Another problem with the left panel in Figure 3.8 is that focusing on one of its lines requires visually separating it from a foreground containing five other lines. With two panels, we can compare two lines by visually separating each line from a simpler plot. If we could highlight our two chosen lines for comparison in red, we could easily compare lines in the new group of size two. Perceptual grouping can be a powerful tool for simplification.

Figure 3.9 illustrates the benefits of perceptual grouping with an interesting data set consisting of brain

and body mass for sixty-two species of mammals, a subset of a larger data set in a study of Allison and Cicchetti (1976). (Our source was Chambers et al. (1983), who cite Weisberg (1980).) Cleveland (1985) produced a dot plot of a subset of twenty-seven of these species. He chose to transform the original data to a single variable, log(brain mass) − 2/3*log(body mass), based on a scatterplot of these data in Carl Sagan's book *The Dragons of Eden* (Sagan 1977). Cleveland's rationale was that mass is a surrogate for volume and that volume raised to the 2/3 power approximates body surface area. Stephen Jay Gould (1979) conjectures that surface area serves as endpoints for many nerves so brain size would scale with surface area, not body weight or volume.

In order to construct a graphic with sixty-two rows for a standard 8.5 × 11 inch page, we need to use smaller fonts and row separation. Readability can be difficult. Some may suggest that sixty-two rows are too many for a single page, but comparisons are easier when all of the items are in view simultaneously. We start our design with a simple list (Figure 3.9, left), ordered from highest to lowest values of this index of relative brain size. Its length is intimidating. On the right, this list is separated into perceptual groups of five animals each.

Man	Man
Rhesus Monkey	Rhesus Monkey
Baboon	Baboon
Chimpanzee	Chimpanzee
Owl Monkey	Owl Monkey
Patas Monkey	
Asian Elephant	Patas Monkey
Vervet	Asian Elephant
Arctic Fox	Vervet
Red Fox	Arctic Fox
Ground Squirrel	Red Fox
Gray Seal	
Roe Deer	Ground Squirrel
African Elephant	Gray Seal
Rock Hyrax: *H. Brucci*	Roe Deer
Raccoon	African Elephant
Galago	Rock Hyrax: *H. Brucci*
Genet	
Donkey	Raccoon
Goat	Galago
Okapi	Genet
Mole Rat	Donkey
Sheep	Goat
Echidna	
Gorilla	Okapi
Cat	Mole Rat
Chinchilla	Sheep
Tree Shrew	Echidna
Gray Wolf	Gorilla
Giraffe	
Horse	Cat
Slow Loris	Chinchilla
Rock Hyrax: *P. Habessinica*	Tree Shrew
Phalanager	Gray Wolf
Tree Hyrax	Giraffe
Jaguar	
Cow	Horse
Eastern American Mole	Slow Loris
Yellow-Bellied Marmot	Rock Hyrax: *P. Habessinica*
Mountain Beaver	Phalanager
African Giant Pouched Rat	Tree Hyrax
Rabbit	
Star Nosed Mole	Jaguar
Arctic Ground Squirrel	Cow
Brazilian Tapir	Eastern American Mole
Pig	Yellow-Bellied Marmot
Little Brown Bat	Mountain Beaver
Guinea Pig	
Giant Armadillo	African Giant Pouched Rat
Kangaroo	Rabbit
Mouse	Star Nosed Mole
Lesser Shoot-Tailed Shrew	Arctic Ground Squirrel
Nine-Banded Armadillo	Brazilian Tapir
Rat	
N. American Opossum	Pig
European Hedgehog	Little Brown Bat
Golden Hamster	Guinea Pig
Big Brown Bat	Giant Armadillo
Desert Hedgehog	Kangaroo
Tenrec	
Musk Shrew	Mouse
Water Opossum	Lesser Shoot-Tailed Shrew
	Nine-Banded Armadillo
	Rat
	N. American Opossum
	European Hedgehog
	Golden Hamster
	Big Brown Bat
	Desert Hedgehog
	Tenrec
	Musk Shrew
	Water Opossum

FIGURE 3.9 The long ordered list on the left is visually intimidating. Using perceptual grouping and color-coded visual entry points helps the reader search the list on the right. Data are an index of relative brain size for selected mammals, in descending order. (Data attributed to Allison and Cicchetti (1976, 194) as listed in Appendix Data Set 19 by Chambers et al. (1983).)

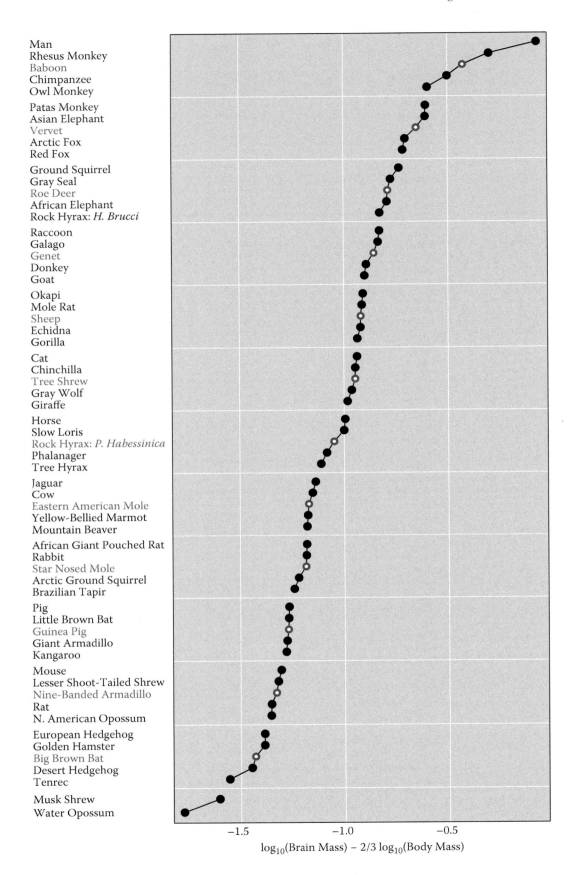

Man
Rhesus Monkey
Baboon
Chimpanzee
Owl Monkey

Patas Monkey
Asian Elephant
Vervet
Arctic Fox
Red Fox

Ground Squirrel
Gray Seal
Roe Deer
African Elephant
Rock Hyrax: *H. Brucci*

Raccoon
Galago
Genet
Donkey
Goat

Okapi
Mole Rat
Sheep
Echidna
Gorilla

Cat
Chinchilla
Tree Shrew
Gray Wolf
Giraffe

Horse
Slow Loris
Rock Hyrax: *P. Habessinica*
Phalanager
Tree Hyrax

Jaguar
Cow
Eastern American Mole
Yellow-Bellied Marmot
Mountain Beaver

African Giant Pouched Rat
Rabbit
Star Nosed Mole
Arctic Ground Squirrel
Brazilian Tapir

Pig
Little Brown Bat
Guinea Pig
Giant Armadillo
Kangaroo

Mouse
Lesser Shoot-Tailed Shrew
Nine-Banded Armadillo
Rat
N. American Opossum

European Hedgehog
Golden Hamster
Big Brown Bat
Desert Hedgehog
Tenrec

Musk Shrew
Water Opossum

-1.5 -1.0 -0.5

$\log_{10}(\text{Brain Mass}) - 2/3 \log_{10}(\text{Body Mass})$

FIGURE 3.10 Adding a dot plot of the Figure 3.9 mammal data visually displays the relative brain sizes for more precise comparisons beyond simple rank order.

Order   Boxplots

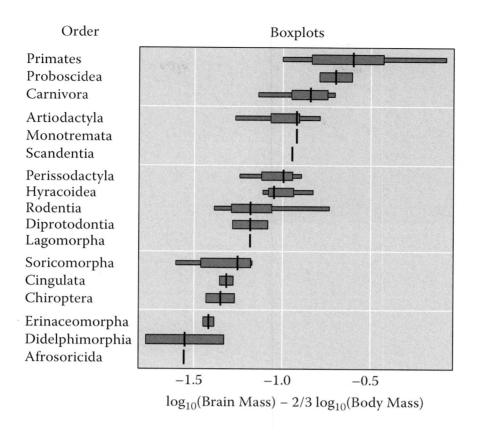

$$\log_{10}(\text{Brain Mass}) - 2/3\,\log_{10}(\text{Body Mass})$$

FIGURE 3.11 Aggregation of the mammals' relative brain weights results in a more compact display but masks details. Data are displayed as box plots, grouped by biological species order.

The middle entry is highlighted in red to serve as a visual entry point, which serves as a focal point in each group and helps us with visual searches of the list. A dot plot of the data (Figure 3.10) allows us to judge not only the rank order of the index but also relative differences among the species. For example, we immediately see that man has a much higher value than the second-ranked rhesus monkey; in contrast, there is very little difference among the mammals in the fifth group.

3.3.2 AGGREGATION

Another method of simplification is to aggregate the data. This is accomplished by means of stratified statistical summaries, the simplest being sums of counts by strata. For example, hexagon binning, defined in Chapter 2, aggregates data within each hexagon-shaped cell on the plot (or map) so that features of very large data sets can be seen. Continuing with our mammal example, we could reduce the number of rows in the graphic to show the distribution of values for mammals within each biological species order by box plots (Figure 3.11). This shortens the list, simplifying its appearance, but masks the identification of members of each group. For

example, the box plot groups most types of monkeys together as primates, so we lose the detail that gorillas are more like cats than chimpanzees (see Figure 3.10). We also have replaced familiar species labels with less familiar order labels, so the plot becomes more difficult to use. This example illustrates the trade-offs of aggregation—the summaries can simplify the appearance of the graphic while retaining important features of the data and can help to mask confidential data of individuals, but also can mask interesting details, such as the variability within the aggregation units.

3.3.3 SMOOTHING

Another method used to simplify the appearance of a graphic is smoothing. A regression line overlaid on a scatterplot is a smooth representation of the relationship between the two graph variables. For time series data, a moving average of the data over time is often used to smooth out the variation over small time steps in order to illustrate the overall trend.

Smoothing maps is more complicated, as we may smooth the boundaries, the data, or the visual encoding method. People associate memories with places and

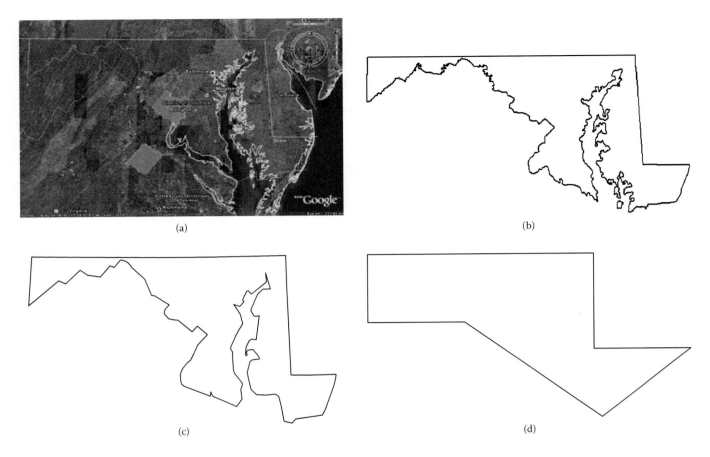

(a)

(b)

(c)

(d)

FIGURE 3.12   State boundary of Maryland (a) as seen from space. (Image from Google Earth,™ Google, Inc. 2009.) (b) Generalized boundary from Census Bureau TIGER files. (c) A smoother version of (b) provided by David Wong. (d) Boundary caricature from Monmonier's visibility map (Monmonier 1993).

regions, so maps often provide helpful cues to recall, but we need enough detail to recognize a place. Two important tasks that maps can serve are to provide recall cues and to provide an organizational framework for storing new memories. The memories can be of data or statistical summaries that we have studied.

The role of maps in micromap plots is to support color linking of values in separate panels to region names, to permit recall of region names from the shape and relative position of the areas, and to provide spatial context by identifying neighboring regions. To paraphrase the Einstein quote opening this section, the map boundaries need to be detailed enough to achieve these goals, but not more detailed. We illustrate boundary smoothing using the state of Maryland. The satellite image in Figure 3.12a shows the detail of the Chesapeake Bay shoreline. For data visualization purposes, generalized (smoothed boundary) maps (Figure 3.12b–d) can convey data patterns just as well. The Census' generalized TIGER boundary file (Figure 3.12b) was further generalized (courtesy of David Wong) for use in

micromap plots that display counties, such as the NCI State Cancer Profiles web-based linked micromaps (Figure 3.12c). Mark Monmonier (1993) produced a state visibility map with very simple and recognizable boundaries, enlarging small states for better visibility, which we modified slightly for our micromap plots of all U.S. states (Figure 3.12d). These generalized maps are faster to draw on the display, and they minimize the boundary ink that can interfere with recognition of the interior color shading for small areas, important advantages for dynamic display of many small maps, the foundation of micromaps.

In addition to smoothing boundaries, we can smooth the data. The simultaneous smoothing of variation over space, time, or attributes can help us to see the central patterns that would otherwise be hidden by local variation (noise). Local averaging of values usually can provide less biased estimates of spatial and temporal processes, just as the regression line can provide an unbiased estimate of a linear relationship between variables. However, smoothing can actually mask patterns, particularly important outliers, if we smooth over places that are dissimilar in some relevant attribute. Neighboring states and countries can be quite different in terms of policy, local environment, health care availability and affordability, etc. As an example, smoothing cervical cancer rates across several adjacent states will make their rates appear more similar, hiding the fact that one state has very high rates due to a lack of screening coverage in their Medicaid program. In addition, some methods consider two areas to be neighbors for smoothing purposes even if they are separated by large distances or by physical barriers such as oceans or mountain ranges. Smoothing works best at highlighting geographic patterns in the data when the regions being averaged together are somewhat similar in the attributes that are important correlates of the study variable.

There are a number of two-dimensional smoothers available (Kafadar 1994), but their suitability for maps depends on the type of data. There are relatively simple local smoothers, such as moving averages and the LOESS (Cleveland 1979) and Headbang (Mungiole, Pickle, and Simonson 1999) algorithms, and more complex parametric smoothers, such as kriging. For higher-dimensional data, Whittaker and Scott (1994) proposed the use of average shifted histograms, where adjacent categories of the conditioning variables are averaged to stabilize the patterns. They demonstrated that the results of multivariate smoothing can be shown in conditioned maps.

Most of these algorithms provide smoothed values for the original geographic units of the input data, while others provide interpolated values for all locations within the study region. In order to smooth data representing administrative areas, such as states, each value must first be assigned to a representative location. Commonly used locations are the geographic or population centroid or the location of the largest city in the area. Interpolation methods, such as spatial regression prediction models, open the door to more mapping options. Instead of encoding the mapped values by shading the geographic units according to a categorical color scheme (a choropleth map style), these methods can predict values on a regular (e.g., square or hexagon) grid. We can then color the grid cells on the map based on the predicted values. This is advantageous for the representation of outcomes that do not respect political boundaries, such as contagious diseases. On the other hand, when outcomes may depend on policies such as coverage of cancer screening by state Medicaid programs, retention of administrative boundaries is important.

Choosing appropriate weights is important, regardless of the smoothing algorithm chosen. The variability of disease rates, for example, depends strongly on the population sizes of the geographic areas from which the rates were derived. If we just smooth a disease rate map using an unweighted algorithm, implying that all rates are equally reliable, we may smooth away important extreme values that are accurately measured. This outcome contradicts the purpose of smoothing, which is to reduce background noise in the data while retaining important patterns. Using a weighted algorithm with weights inversely proportional to the variance or standard deviation of each rate will retain the values of places with reliably estimated rates (i.e., places with large populations) while modifying less reliably estimated rates to be more like rates of neighboring areas.

We illustrate this using HIV mortality rates in Figure 3.13. In general, HIV rates are high in urban areas and low in the surrounding suburbs. By smoothing without weights, the high city rates (e.g., Seattle, Chicago, Minneapolis, etc.) are smoothed away even though their original rates were very reliably estimated (compare Figure 3.13a and b). Weighting by population, the high city rates are retained but rates in less populous places are smoothed to be more like their neighbors' (Figure 3.13c).

As you can see by comparing the maps in Figure 3.13, smoothing the values to display increases the amount of spatial autocorrelation in the mapped data. That is, any smoothing algorithm makes each region look more like its neighbors. This is an advantage in that spatial patterns are strengthened and thus easier to see, but a disadvantage if you want to apply traditional statistical methods to these data later. For example, in Chapter 5 we will describe some simple statistics that can help us to judge the degree of association between the study variable and conditioning variables. These methods should be applied to unsmoothed data because they do not account for any spatial autocorrelation in the data. Models for autocorrelated data are beyond the scope of this book; several excellent books are now available on this topic (Banerjee, Carlin, and Gelfand 2004; Schabenberger and Gotway 2005; Waller and Gotway 2004).

Sometimes we wish to use a display method that results in a smoother visual interpretation of the map. A second mapping option enabled by spatial models or other interpolation algorithms is the production of isopleth maps, which connect by lines any locations that have approximately the same value. Daily weather maps

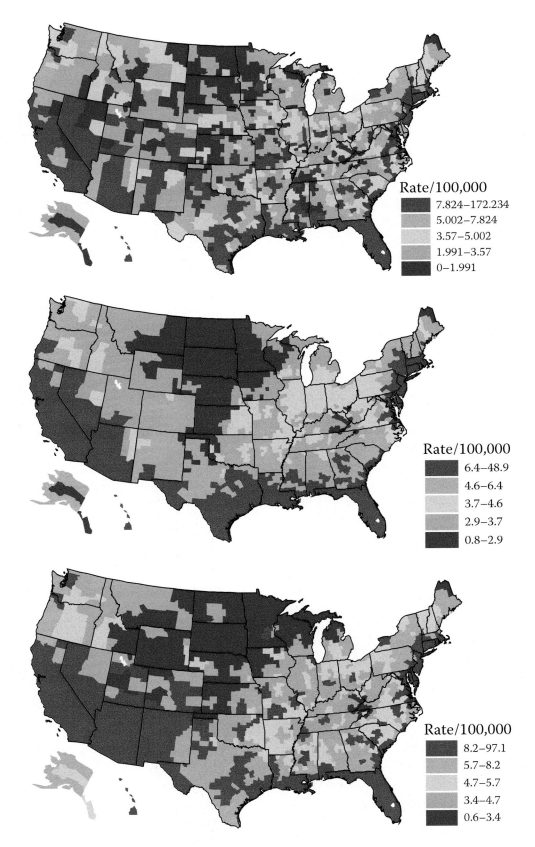

FIGURE 3.13 An example of spatial smoothing. Rates of death due to HIV infection among white males, 1988–1992 (top) are smoothed by the Headbang algorithm without weights (middle) and with weights proportional to population (bottom). High city rates are retained by using weights inversely proportional to rate variances or standard errors.

of temperature are often isopleth maps. Similar to gridded maps, the isopleth map removes the dependence of the map pattern on administrative boundaries.

This map style has not been implemented in micromaps, but it could be. Instead, some micromap applications can display data aggregated to a hexagon grid, which also displays patterns independent of political boundaries.

To summarize our discussion of smoothing, micromaps use smoothed boundaries and can display smoothed data either as input by the analyst, as aggregated to hexagon bins, or as calculated by a statistical model. One important application of micromaps is to show both smoothed values and residuals (differences between the original and these smoothed values).

3.4 ENGAGE THE ANALYST

Engaging the analyst, i.e., attracting and keeping his or her attention, is a nontrivial task. Analysts range from the casual readers who become curious about inconsistencies, puzzles, and mysteries to powerful thinkers who derive satisfaction from flexing their intellectual muscles to learn and discover. Quantitative graphics for analysts need to meet them part way, to provide enough sense of value added that it is worth their time to begin thinking, to begin retrieving their relevant knowledge, and to begin considering what they see from multiple perspectives.

Promoting serious engagement requires providing transformational choices, feedback and guidance for taking next steps, pathways to further education, opportunities to share expertise in like-minded communities, and visual analysis management. In this section we describe some easily implemented features that can help.

3.4.1 ATTRACTING THE ANALYST

Since most producers of quantitative graphics don't have advertising budgets to use for attracting readers, we assume there must be some interest or motivation that leads a potential analyst to at least glance at a graph. Thus, an important goal to draw in moderately motivated analysts is to survive the glance test. We distinguish two evaluative categories, complex and boring, that drive people away.

With a twinkle in his eye, Carr (Carr and Olsen 1996) proposed a complexity measure for graphics or tables called the visual intimidation factor (VIF). The shortened form was intended to rhyme with whiff,

which can be imagined as the sound of the air when attempting to strike an object, such as a ball with a bat or racket, and totally missing. In other words, to VIF means to totally miss the target. The measurement proposed was 1 divided by the length of time in seconds to decide that a table or plot was not worth using. If a plot seems worth the effort, the VIF is 0 ($1/\infty$). Since it takes us about ½ second to become aware of what our eyes have decided, an upper bound of the VIF is 2.

Others have attempted to quantify the effort required to understand cognitive information economics. Pirolli (2003) describes resource costs as expenditures of time and cognitive effort to obtain information, and opportunity costs as the loss of benefits that could have been gained from other activities. We are more interested in cognitive analytic economics, where the most primitive task is to make meaningful comparisons, and other sense-making tasks range from finding patterns to building empirical models.

What are some simple design features that attract the eye? Boring plots will not attract the analyst. Neither will overly complex-looking plots (with high VIF values). Simplicity and the unexpected grab people's attention (Heath and Heath 2007). Color, particularly large blocks of very saturated colors, draws the eye, as does motion if applicable. The availability of tools that allow the reader to manipulate the graphic and to view the data from different perspectives is further enticement; such tools are not available in static plots. Providing visual entry points, especially for long lists or otherwise complex graphics, can guide the reader to useful starting points. The perceptual grouping of rows can provide visual entry points—recall the mammal list in Figure 3.9. Highlighting some extreme value or interesting comparison can get the reader to take a look. Dan recalls a conversation he had with a food service worker who once had compared putting out bowls of corn with and without a bit of pimiento added. He said students picked the latter because it had the appearance of value added. Adding value to our graphic can attract the analyst, as long as what we add does not distract from the main data story.

3.4.2 ENCOURAGING CONTINUED USE

Many people who look for patterns in data are analysts seeking knowledge.

Their minds are active. They are ready to ask questions, spot patterns, generate hypotheses about patterns, and sometimes willing to pursue their questions and hypotheses. The pursuit may include looking at the immediately available data in different ways, obtaining additional related data, accessing literature and experts for further education, advice, and guidance, and sharing answers and continuing questions with colleagues and interested communities. Active analysts seek data and visual analytic tools. How can we encourage continued use of the graphic after the analyst has once glanced at it?

First, the graphic should appear interesting and any interactive tools should be easy to use. A reader will want to continue exploring the data with the graphic if it is not overly complex and especially if it tells a credible and interesting story about the data (Heath and Heath 2007). In his book on visualizing data, Cleveland (1993) says, "Tools matter." We would expand his powerful and succinct statement to "tools and choices matter." Most of us are capable of visual thinking and enjoy solving problems and communicating with simple visual tools (Roam 2008). Such tools encourage more data exploration and can stimulate our imagination and creativity, helpful for hypothesis generation and general problem solving. We discuss several helpful tools and available options in the next section.

The reader's interest will more likely be maintained if the data stories can be told from the user's point of view. Egocentric graphical views are drawn from the user's perspective—think of the amusing maps where your town predominates in the foreground and all other places are compressed together in the distance. Our mental maps are usually like this, distorting distances so that nearby places are closer than they really are. Using egocentric maps and familiar geographic units in our design facilitates usability by reducing the need for an initial orientation step in understanding, i.e., "What am I looking at?" We can provide views from alternative perspectives that address the hierarchical goals of the analyst. For example, the ultimate goal may be to eradicate cancer, but a series of stepwise, more achievable goals might be: identify and monitor risk factors, detect and understand cancer and risk factor patterns, identify and eliminate cancer disparities of detection, and treatment. In order to address these various goals, we can provide views from various space, time, attribute, internal comparison, relationship, and scale (STAIRS) perspectives.

3.4.2.1 Space

Tobler's (1970) first law of geography states: "Everything is related to everything else, but near things are more related than distant things." This is true for many attributes, so offering the option to map the data is most important. However, it is important to remember that the variation across political boundaries, e.g., between neighboring nations, or across physical barriers such as rivers or mountains, can be so large that maps are not very helpful.

3.4.2.2 Time

Similarly, if the data vary over time, it is important to offer the option of some type of temporal display. We have discussed the merits of simultaneous display versus animation, but slider bars or other tools can be used to restrict the time period to short ranges for data exploration.

3.4.2.3 Attributes

It is often helpful to first look at geospatially indexed data in scatterplots without the complication of geographic context. In Hans Rosling's Gapminder World web application (Gapminder Foundation 2009), data for many nations of the world are displayed in scatterplots with population size represented by circle size and continent of the country represented by color fill. That is, four attribute values are displayed for each nation—two by position on the scatterplot and two by symbol characteristics. The graph can be animated over time, a fifth attribute, with animation trails to help the reader see the time trends. The symbol encodings are poor for continuous variables because of the difficulty in accurately perceiving the size and the color of the circles. However, the graphing tool is easy to use and the analyst may choose from many attributes and examine relationships among them.

3.4.2.4 Internal Comparisons and Relationships

Comparisons of the data with external referent data or internal comparisons will usually put the data values in context and help with interpretation. Examples of internal comparisons include the calculation of standardized ratios, such as observed disease rates for each place relative to expected rates based on the total of all places, and percent of a total. Box plots show internal variation within the larger unit, e.g., county rates within each state. An illustrative external comparison is the addition of a reference line to a linked micromap plot showing the goal for the plotted values, such as the Healthy People 2010 goals for lowering disease rates and risky behaviors (U.S. Department of Health and Human Services 2000). Relationships among the attribute variables are most often examined by pairwise scatterplots, but other display methods are available, such as parallel coordinate plots (Wegman 1990) and linking plots to maps by color (White and Sifneos 2002).

3.4.2.5 Scale

Because patterns can vary by scale, it is important to allow the analyst to examine the data at several different scales. This can be accomplished by a choice from a short list of available data scales, such as county or state, or a zoom capability. Good examples of this sort of choice are the CDC BRFSS maps (Centers for Disease Control and Prevention 2008), which show data for state and metropolitan areas, and the NCI State Cancer Profiles maps (National Cancer Institute 2006), where the analyst can choose a national micromap of states or a state micromap of counties.

3.4.3 EMPOWERING THE ANALYST

The very best static plot that we produce may answer one or more questions and may raise others, but static plots are inherently limited. Static plots cannot anticipate all the analyst's questions and can only support a few ways to think about data. If we are to engage analysts, we must provide them with choices in tools that can instantiate those choices.

3.4.3.1 Analytic Management Tools

For displays from any of these (STAIRS) perspectives, there are data and visualization management tools that can enhance the analyst's experience. For example, we can allow the option of different transformations and models of the data. Rosling's Gapminder plot allows the analyst to choose between a linear and logarithmic representation of the data on each axis. Supplemental statistics, statistical tests, or distributional summaries are often helpful in guiding the analyst. Examples include conditional or marginal plots; regression lines, correlation coefficients, or goodness-of-fit statistics for scatterplots; or a cluster detection tool for maps. Some people will not have the quantitative skills to use the more sophisticated tools, so the defaults should be chosen that will be understandable by all, and the options offered should be compatible with the quantitative skills of the expected audience.

When more options are available to the analyst, the path of data exploration can become complicated. Allowing the user to annotate the displays, by either text or audio, and saving program parameters of favorite views are important options that make the analysis replicable. This also allows the analyst to restart at key places by saving the state of the system. The saved information can also be edited to restart the analysis with different options or different data or to resequence the views for live presentation later.

3.4.3.2 ANALYST SUPPORT

Despite our best design efforts, analysts will have questions about the use of the visualization tool and its options. A good help system is important. A tutorial can help the new user begin to use the tools quickly and effectively, reducing the frustration of learning a new system that can discourage continued use of the tool. Other learning pathways may be useful, such as worked examples using constructed data sets.

To illustrate how the designer can go beyond a single help button with a list of explanations, Figure 3.14 is a downloadable quick reference guide for linked micromaps as implemented in the NCI State Cancer Profiles system. This is meant to be printed out as a reference for the user in addition to the usual help screens. Each type of output from this website has a corresponding guide. The informational boxes explain the meaning of a symbol or instruct the analyst how to use a particular feature, such as "Click on a triangle to sort by this column." Note that "Source" is displayed at the bottom left of the screen; clicking on this will display the detailed data source.

Finally, if the results of the data exploration are to be presented, especially in a scientific publication, output in high quality format should be available. This is often different than an exploratory format. In addition, the source of the data, if provided by the graphic designer, must be clearly identified. If at all possible, permitting access to the original data either by direct download or by a link to the data source encourages further analysis offline. For example, the NCI State Cancer Profiles and mortality atlas online systems cannot provide case-specific information because of privacy regulations, but they do provide stratified cancer count and rate data as displayed. Map boundary files used for the atlas maps are also downloadable from the NCI website to facilitate mapping of other data by the analyst (www3.cancer.gov/atlasplus).

Designing good visual displays with an easy-to-use interactive system is difficult. The designer's first attempts will usually fail, so it is critical that proposed systems be tested on at least several sets of typical users. These usability tests help the designer iterate to the best possible system. For general advice about developing website designs, see the guidelines developed at the

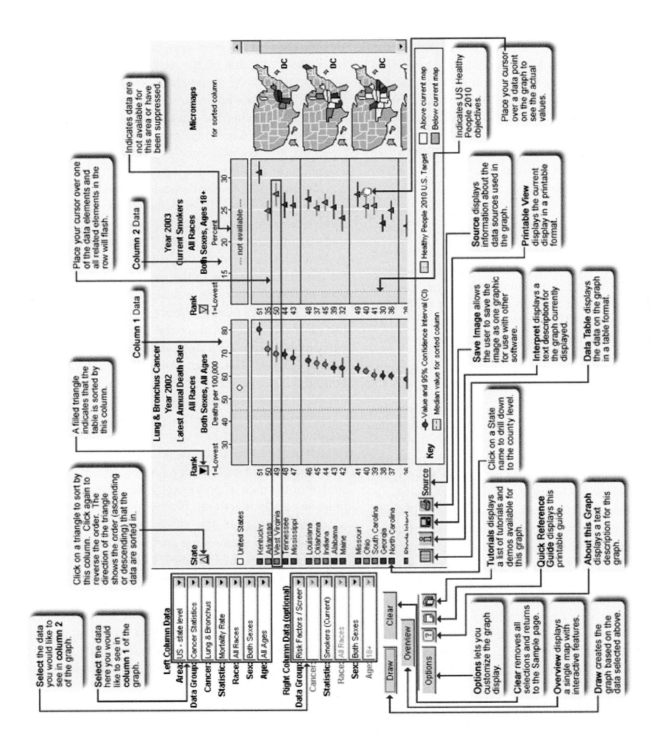

FIGURE 3.14   Quick reference guide for linked micromap plots as implemented in the National Cancer Institute's State Cancer Profiles website (http://statecancerprofiles.cancer.gov).

TABLE 3.1
Examples of Contextual Questions for the Data

Who	Who is providing the data (original source)?
	Who has vested interests in this summary?
	Who can provide more information?
What	What is being displayed?
	What was the process that generated these data?
	What are the conclusions and what actions are to be taken?
	What are the units of measure, e.g., are rates per one thousand or per one hundred thousand population?
	What potentially important features have not been measured?
When	When were the measurements made? Were these specific times or intervals?
	When were the data processed?
	When were the processed data reported or released?
Where	Where were the entities or processes located, e.g., state or region names?
Why	Why was the data collected?
	Why was this study conducted?
How	How were attributes computed?
	How were original data recoded, e.g., for missing values?
	How were variables transformed, e.g., how were rates age adjusted?
	How were missing data treated?
	How was the original data collected, e.g., sampling, experimental design, or survey?
	What analysis methods and assumptions were involved?
	How can I use this tool?
How well	How representative and relevant is this summary to the population or phenomenon of interest?
	What are the error rates of data collection or reporting?

Department of Health and Human Services, available online at www.usability.gov.

3.4.3.3 Promote Meaningful Interpretation

Providing context is an important part of telling a story. As we mentioned at the beginning of our tour in Chapter 1, the context for quantitative stories often includes information about who, what, where, when, why, how, and how well. Sample questions are shown in Table 3.1. The statistical graphic only tells part of the story.

The analyst cannot correctly interpret the visualization patterns if he or she does not understand the data or process by which the visual summaries were created. Most of the information listed above is delivered via help screens or in titles or footnotes on the display itself. Tutorials can lead the analyst through typical exploratory paths, generating ideas for how he or she might use the tools for his or her own data or for different subsets of provided data. Note the i symbol at the lower left in the State Cancer Profiles help sheet (Figure 3.14), which brings up text that interprets the graphic as displayed. This is helpful to the analyst who is new to the tool, who learns more from reading than from visual patterns, or who wants to know how the agency experts interpret these same data.

As more and more information is provided over the web, sharing data, analytic tools, and visualizations of data are becoming more prevalent. It is going to take a community effort to develop serious visual analytic tools. Increasingly powerful statistical (www.statlib.org), mathematical, and mapping scripting resources such as R (R Development Core Team 2009), Python® (www.python.org; The PythonSoftware Foundation), and ArcGIS® (http://arcscripts.esri.com; Environmental Systems Research Institute 2003) have shown the merits of global community contributions. Geographic information systems (GIS) have become ubiquitous—everyone can make a map on their computer now. Layers of information from one analyst can be melded together with that from another to produce a map richer in content than either of the original data maps.

There is merit to having the input of an informed community at the early stages of data exploration, prior to publication. An interactive discussion about the patterns on the graphic and their interpretation can uncover alternative opinions about the reasons for the patterns or point out flaws in data collection or interpretation. Email listservs also can serve this purpose, as an analyst can send a result or graphic to the group for comment. An IBM Visual Communication Lab project called Many Eyes promotes a "new social kind of data analysis" (http://manyeyes.alphaworks.ibm.com/manyeyes/page/About.html) by encouraging comments (blogging) about contributed graphics. Multiple viewers often can provide a clearer explanation of a pattern or related knowledge about the subject.

3.5   SUMMARY

In this chapter we have applied the research summarized in Chapter 2 to develop recommendations for effective visual design. The most important design goal is to enable accurate comparisons, since comparisons are at the heart of micromaps. We can view data from many perspectives so comparisons can take many forms. The acronym STAIRS can jog our memory about the six common classes of comparisons:

S: Comparisons in a **spatial** context
T: Comparisons in a **temporal** context
A: Direct comparison of **attribute** values to each other or to global reference values
I: **Internal** comparison when values are summaries of parts
R: Comparisons that show **relationships** among attributes
S: Comparisons at different **scales**

The graphic designer often has several sound designs to choose from. We argue that the best choice among these is the simplest visual display and the design that will attract potential analysts and allow them to interact with the display to achieve their goals. Finally, we need to support the analyst by providing background material, help screens, and tutorials for new users and information that aids in the correct interpretation of the data. To design successfully, we need to know our audience—their goals and technical skill level—and to test our proposed designs on representatives of this audience. Only then will we know whether we are close to that elusive goal of an optimum design.

4 Linked Micromaps

4.1 INTRODUCTION

In Chapter 1 we introduced linked micromaps, graphics that link statistical graphics to an organized set of small maps. As we saw in Chapter 1, the maps need not show location on the earth but can display any two-dimensional location, such as positions of baseball players or locations of parts of the body. Linked micromaps have been implemented by individual analysts for specific data explorations (Gebreab et al. 2008) and by the National Cancer Institute to explore and communicate cancer statistics. We hope to inspire you to think of how they could help you to explore and present your own data.

As noted in Chapter 1, Dan sought to promote the use of statistical graphics by federal agencies by promoting graphic designs such as dot plots and by showing how tables could be converted to plots. He suggested that digital media could serve archival and data access purposes better than printed tables and that statistical graphics would serve federal communication needs better than tables. The technical report *Converting Tables to Plots* showed plots that resulted from converting data from EPA tables (Carr 1994). During a research visit to EPA–Corvallis, Tony Olsen, a key collaborator, asked if there might be some way to combine row-labeled plots with maps. This question sparked Dan's flash of insight now known as linked micromap plots.

The first paper with linked micromap examples appeared in 1996 (Carr and Pierson 1996), and the label linked micromap plots appeared in 1998 (Carr et al. 1998a). The graphic elements that came together in linked micromaps had been around for some time. For example, Tufte (1983) had long been an advocate for the use of small multiples and parallelism, as had Bertin (1973) in earlier work. Cleveland's dot plots were an important precursor (Cleveland and McGill 1984). Previous research had demonstrated the use of color to link graphics. An interesting early example linked two subsets of acid monitoring sites in maps (point-in-polygon selection defined the subsets) with a hexagon binned scatterplot matrix showing bivariate density differences for several water quality attributes (Carr et al. 1987). Monmonier (1988) had addressed juxtaposing maps and statistical graphics.

The creation of linked micromaps was both part of the spirit of the times and the result of a pressing need to map multivariate environmental statistics.

Later enhancements of the micromap design drew on cognitive science findings, such as perceptual grouping and layering information (Carr and Sun 1999). Applications of these early ideas can be seen in the National Center for Health Statistics (NCHS) atlas (Pickle et al. 1996), where Carr proposed the color link between a regional graph and a small area map, and in a National Cancer Institute (NCI) website to disseminate cancer statistics (statecancerprofiles.cancer.gov), where he worked directly with NCI staff to tailor the linked micromap design for web use.

The basic layout of a linked micromap consists of a column of graphic panels for each plot variable linked to a column of small maps by color. In the simple example in Figure 4.1, recent cancer mortality rates for twelve states in the Midwest are shown. The observed rates are represented by color-filled circles and the corresponding 95% confidence intervals are shown as lines in the same color. The states are sorted from lowest to highest rates. We can see by reading left to right that North Dakota is colored red, that it has the lowest rate of these twelve states, although it has a wide confidence interval, and where it is on the map. Note that the color is used solely to link the elements of the micromap: area labels, micromaps, and plotted statistics. Also note that the map has very smoothed boundaries but you should have no trouble recognizing these states. The two replications of this initial linked micromap plot identify the column and row perceptual groups.

While linked micromaps do not scale to showing thousands of places in a single view, they can be very good for exploring and telling stories about a modest number of regions. A single printed page can comfortably accommodate about fifty rows of information. In this simple example, we see a clear geographic trend of cancer mortality rates among white men in this region. Why is that? We will see from further examples that we can append columns of statistical summaries for other regional attributes in order to examine the association between these factors and disease rates.

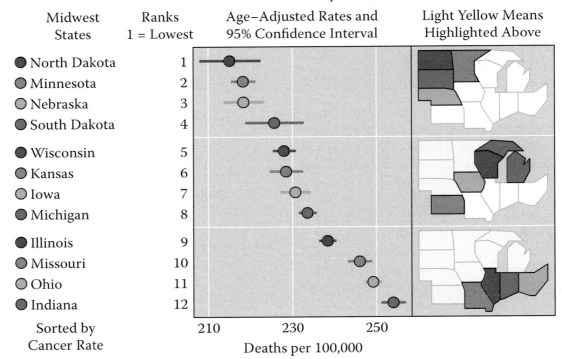

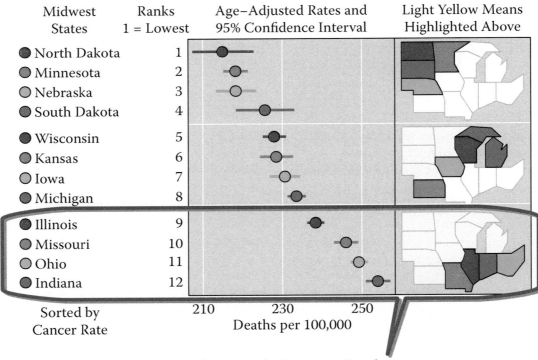

A Perceptual Group of 4 Rows in 4 Panels
(names, ranks, statistics, micromaps)

FIGURE 4.1 An example illustrating the basic components of a linked micromap plot. The top plot, showing age-adjusted rates of mortality due to any type of cancer among white males, 2001–2005, for twelve of the Midwest states, is repeated below with annotation.

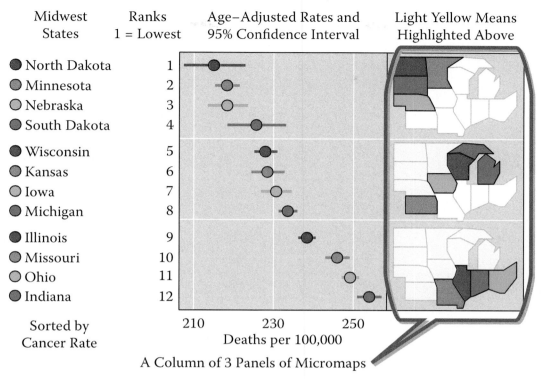

White Male All Cancer Mortality Rate, 2001–2005

A Column of 3 Panels of Micromaps

FIGURE 4.1 (CONTINUED)

In this chapter we explore variations and enhancements of this versatile design. We start by discussing the overall page layout, and then examine methods useful for encoding data on the maps and graphics. We follow these basics with enhancements, a discussion of available software, and future directions.

4.2 PAGE LAYOUT

4.2.1 PERCEPTUAL GROUPING OF ROWS

The linked micromap in Figure 4.1 only displays data for twelve states. As there are more and more geographic regions to plot, the micromap can become visually intimidating and difficult to use. In this section we discuss perceptual grouping and layout variations that can help simplify appearance. The states in Figure 4.2 (reproduced from Figure 1.3) are sorted by the percent of households with income below the poverty level in 2000. The perceptual grouping in the layout partitions the states into groups of four, following recommendations from Kosslyn (1994a) and others in human perception and cognition as summarized in Chapter 2. We can easily compare values and find places on the map when we focus on a single row of panels with only a few geographic units.

There are seventeen states in this linked micromap, not a number evenly divisible by 4, so there is one state left over if we fill all the other groups. How should the singleton group be handled? When considering this situation, where the number of regions is not evenly divisible by a desired group size, we might first think to put the partial group at the bottom of the plot. However, design elegance often includes symmetry, so we chose to group states in the 4-4-1-4-4 pattern, as illustrated. An advantage to this design is that it calls attention to the state with the median household poverty level, which in this example is Oklahoma.

Taking this one step further, we can define a hierarchical grouping of groups. Grouping by size and putting the single member group in the middle yields three groups in a 4-4 | 1 | 4-4 pattern, where the vertical line separates adjacent groups of panels. Figure 4.2 has a black outline around the whole middle row of all the panels for the single member group to suggest the division into three groups. The top and bottom groups of data panels are also outlined in black to emphasize this higher order of grouping.

Figure 4.2 is a special design that omits the micromap for a middle group of size 1. This saves vertical space, because the micromaps use more vertical space

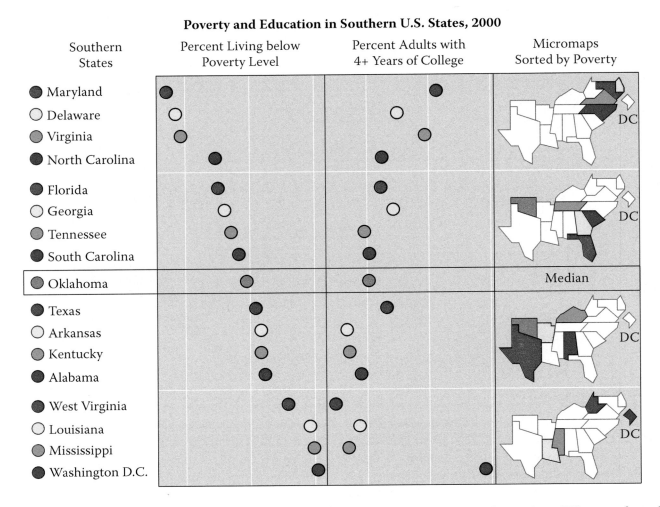

FIGURE 4.2 Linked micromap plot of two variables—college education and poverty—for southern U.S. states from the 2000 Census. The state with the median value of the sort variable (poverty) is highlighted, but without a corresponding map.

than needed for a row of panel labels and data panels for a single state. A fifth color, gray, is used to link Oklahoma to its dots and the polygons in the micromaps above and below the omitted micromap. Vertical space may not matter much for this example but becomes important when showing all the states on one page, as we illustrate next.

The data from Figure 4.2 are expanded to the entire United States in Figure 4.3 using the perceptual grouping for fifty-one states developed by Carr for use by several U.S. federal agencies. Agencies are required to include the District of Columbia with the fifty U.S. states even though in many respects it is more like major U.S. cities than like U.S. states. Black outlines and white space in Figure 4.3 encourage us to focus on three major groups, a block with five rows of panels at the top, each panel displaying five states; a block with five rows of panels at

the bottom; and a block with one thin row of panels in the middle. This higher-order grouping of size three is cognitively simple. When we focus on the top block, we can examine more closely one of the five rows of panels. Then we can focus on the five states in a particular row of panels. Comparison of Figures 4.2 and 4.3 suggests that there is still an inverse correlation between poverty and education across all states, but that this correlation is not quite as strong as we saw among only southern states.

Kosslyn indicates four is the best number of items in a perceptual group, so why group by fives? Working with groups of five is a compromise that we often make for two purposes. The first is to avoid long intimidating lists of higher levels of grouping. Grouping fifty-one states by four produces a 4-4-4-4-4-4-3-4-4-4-4-4-4 symmetric pattern with thirteen groups. Even this long

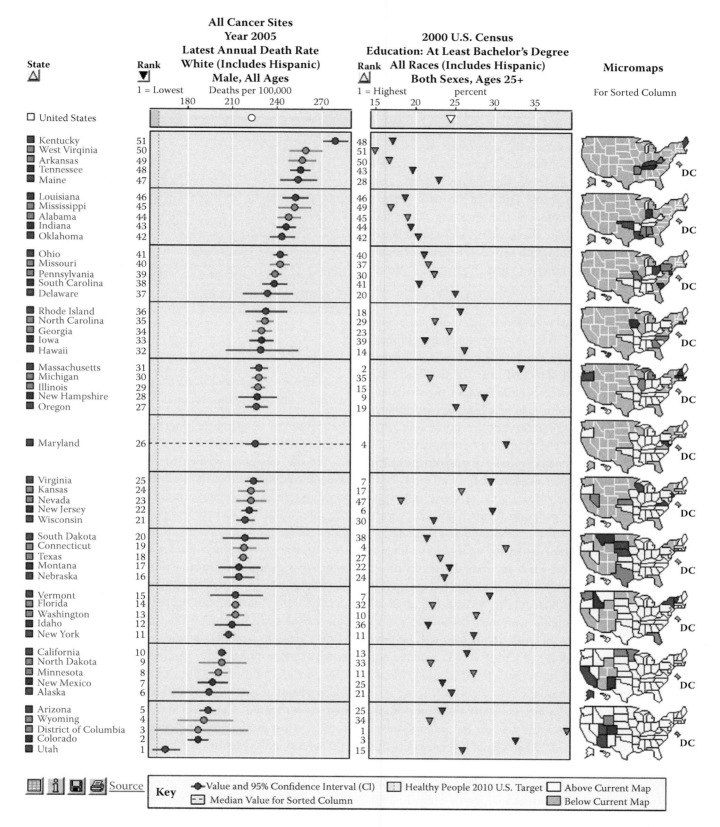

FIGURE 4.5 NCI State Cancer Profiles web implementation of linked micromaps with added features. Data are age-adjusted rates of mortality due to any type of cancer among white males, 2005. (National Cancer Institute 2008b.)

degree of uncertainty in the graph values is understood. If the initial micromap display does not include confidence intervals, the reader may be "anchored" to the patterns seen in the point estimates even after the confidence intervals are turned on. That is, readers may still believe that the ranks and comparisons of the points are exact without an understanding of the varying uncertainty of the estimates across the geographic units.

The absence of confidence intervals on a graph should raise questions. How statistically sound was the data collection process? Did that process permit calculation of the variation of the resulting statistical summaries? For example, the U.S. EPA Toxic Release Inventory reports on company self-assessments of their toxic environmental releases, and historically these estimates have appeared without confidence intervals (www.epa.gov/tri).

Linked micromap confidence intervals are constructed in the same way as for any other statistical graphic. For normally distributed data, $\bar{y} \pm 1.96 \; s/\sqrt{n}$, where $s/\sqrt{n}$ is the estimated standard error of the observed mean $\bar{y}$, encompasses the middle 95% of the distribution of values. Other widths can be used, but we prefer to use the conventional 95% interval to be consistent with readers' expectations. The frequentist interpretation is that if we were to repeatedly collect random samples from a population and calculate the mean and 95% confidence interval each time, 95% of these intervals will include the true population mean. Thus, the width of the confidence interval can be viewed as a measure of the reliability of the point estimate. Visually, the longer lines associated with the least accurate estimates draw the eye more than the shorter lines associated with very accurate estimates. This seems to violate our goal of focusing the reader's attention on what is most accurate, but without confidence intervals the reader will believe that all values are equally reliable. This is usually not the case, as for disease rates that are more reliably estimated in larger populations.

Because the point value is usually more important than the confidence interval, we plot the point symbol on top of the interval line. This visual layering puts the interval line more in the background so that the symbol stands out. A disadvantage of this order of layers is that a short confidence interval can be completely hidden behind the symbol, leading the reader to wonder if there is any interval displayed at all. The point value can be emphasized even more by using a large symbol size, by outlining the symbol in black, by using a thinner line for the confidence interval, or all of the above. We use the

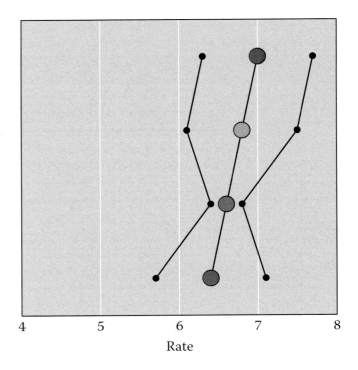

4   5   6   7   8
Rate

FIGURE 4.6 Example of a single linked micromap panel with vertical connection of the endpoints of confidence intervals.

convention of coloring the interval line the same as the point symbol.

If the designer wishes to draw less attention to wide confidence intervals, i.e., to draw more attention to more reliable estimates, an alternative is to represent the interval as dots at the endpoints instead of a line. Connecting the dots downward across the rows gives the impression of pinch points where the confidence interval is unusually short. Even when the interval is shorter than the width of the point symbol, the presence of connecting lines implies the presence of a hidden confidence interval. Examples like that in Carr and Pierson (1996; a single panel is shown in Figure 4.6) have met with mixed response. The arguments for connecting the dots include guiding the eyes to the next points, creating perceptual groups, and creating simple shapes that can be compared (Carr and Sun 1999). Arguments put forth against connecting the dots include that it is not a conventional graphing method and concern that this might imply interpolation, although we fail to see how a reader would think of interpolating between nonadjacent states or between qualitative categories such as man and rhesus monkeys (Figure 3.10). The pinch point idea is still alive—Symanzik used this method in an exploration of birth defect data (Gebreab et al. 2008). Experiments are needed to evaluate the merits of this method before widespread adoption, but if the intent is to draw attention to the very reliable values on the micromap, it seems to work well.

One disadvantage of using arrows as the basic data encoding is the difficulty of adding confidence intervals to the endpoints. It may be possible to add confidence lines close to the arrow, say, below the first endpoint and above the arrowhead without much clutter or confusion. This is speculative, but gives an idea to a designer who wishes to try to include reliability information with the arrows.

Confidence intervals display the variability of the plotted estimate, but if a summary statistic is plotted, the shape and variability of the underlying data distribution is masked. A small confidence interval for a plotted estimate indicates a precise estimate at that scale, but not necessarily at a finer scale. For example, a state's age-adjusted cancer mortality rate may be very reliably estimated because of the large state population, but some county rates could be very different from the state rate.

A **box plot** summarizes the shape of a distribution, identifies outliers, and can be used on masses of data. Components of the box plot are defined in Figure 4.7. Visually, the length of the boxes for several subsets of data can be compared to instantly identify which subset has more or less variability than the others. Because this graphic reduces the original data set to a five-value summary plus outliers, it works equally well on large and small data sets. Although the diagram may appear complex, a version of the box plot is now being taught in some elementary schools.

Supplementing a column of age-adjusted rates (or means) and confidence intervals for states in a linked micromap with a column of box plots of county values shows the rates at two different geographic scales, as shown in Figure 4.8. The box plots convey the geospatial distributions of county rates within each state without identifying the county locations and can provide a basis for drilling down to counties with unusual rates. Neither the State Cancer Profiles website application nor the NCI quality control (QC) linked micromap software produces this combination of plots yet, but we provide R code on the website to implement the example shown in Figure 4.8.

Let's unpack the information in this figure to see how much can be shown on a single page. There is over a twofold difference in rates for 2000–2004 from highest to lowest state. The confidence intervals are only visible for states with relatively small populations; intervals for more populous states are shorter than the diameter of the dots, indicating a highly reliable rate estimate. In the

second column the short arrows indicate that there has been little change for most states during the five-year period, as would be expected. Of concern is the fact that ten of the top eleven states have rates that have increased since 1995–1999. Delaware's rates rounded to the same value for both periods, so this value is shown as a dot, not an arrow. Reading down the micromap column, we can see a cluster of high-rate states expanding along the Mississippi River. The box plots indicate substantial variability of county rates within some states. For example, at least one county in Nevada, the highest-rate state, has a rate below the median rate (Montana), and a county in Utah, the lowest-rate state, has an above-median rate. This plot would probably lead a cancer researcher to drill down to states of interest to see what the within-state patterns are, as one can do with the State Cancer Profiles web application.

Many people like to think in terms of **ranks**. What is the rank of my state? Is the local sports team ranked number 1 this week? This inclination to think in terms of rankings motivates adding a narrow column of ranks to a linked micromap. We usually choose to just print the numbers, without panel outlines or color links. The horizontal alignment and perceptual grouping gaps should suffice to keep horizontal linkage errors small. Figure 4.5 provides an example with ranks shown for each column. Ranking can be pretty straightforward, but there are two basic issues to address: different ranking conventions and the treatment of ties.

There are different conventions concerning how cases are ranked. Does number 1 mean first or lowest? In sports, teams are often listed in rank order with the best team ranked first based on the highest win-lose percentage, the highest number of poll votes, or another criterion. Thus, conventional practice creates an association between rank 1 and the highest score, but this practice does not generalize to many other situations. Even in sports statistics, the ranking can be reversed so that the best player or team is the one with the lowest statistical value, often for defensive aspects of the game. For example, the baseball pitcher with the lowest earned run average will be ranked number 1 and a team's defense will be ranked first if it has the fewest points scored against it.

There is clear communication benefit in associating number 1 with being the best. The NCAA basketball tournament starts with sixty-four teams. If the winner were simply stated as having rank 64 at the end of the tournament, people would have to know that there were sixty-four teams in order to translate this as best. How

Box Plot Construction

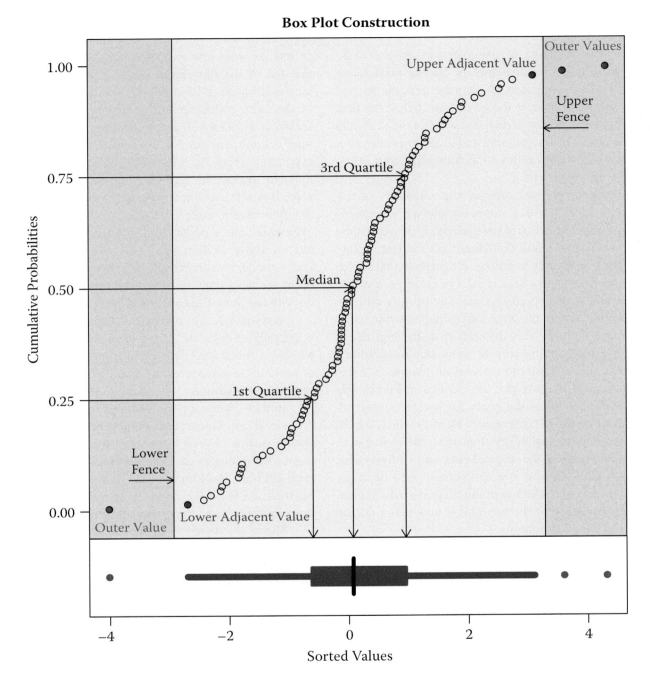

FIGURE 4.7 Definition of components of a box plot. Lower fence = first quartile − 1.5*(third quartile − first quartile). Upper fence = third quartile + 1.5*(third quartile − first quartile).

many teams are in the world soccer cup tournament? Casual enthusiasts may not know. Using number 1 to indicate the best is easier in terms of communication.

What is best can be associated with the smallest value, the largest value, or even a value in the middle, as is the case for a desirable body weight. The State Cancer Profiles website application assigns rank 1 to the best value, even if this causes an inconsistency of rank assignments across columns of the micromap plot. Figure 4.5,

created by this web program, displays default ranks of 1 for the lowest mortality rate and for the highest education rate and labels each column with this definition. The designers made the decision to use this ranking definition after much discussion and usability testing. That is, in this system a rank of 1 indicates "best," but the analyst has the option to change this. For consistent communication in this book, we use the statistical convention of associating rank 1 with the smallest value of

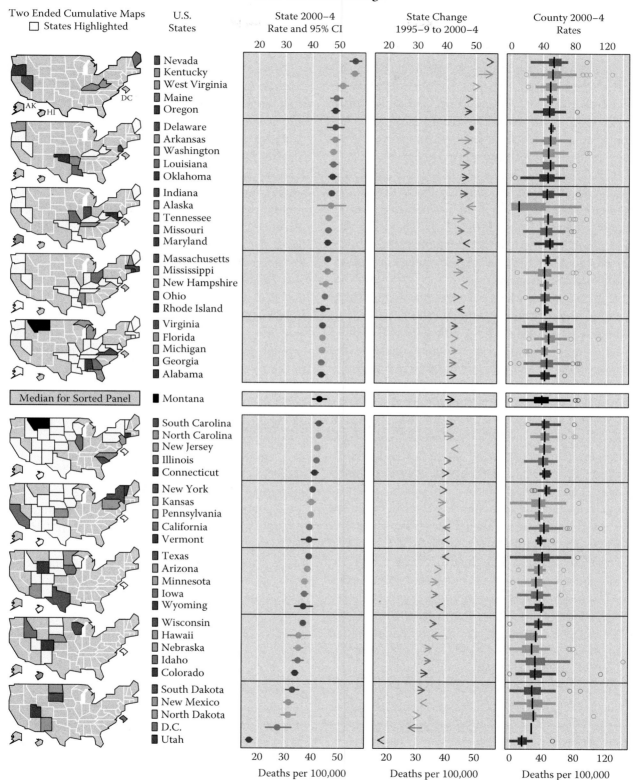

FIGURE 4.8 Linked micromap plot of average annual female lung cancer mortality rates plus confidence intervals in 2000–2004 by state. Rate changes between 1995–1999 and 2000–2004 are shown as arrows, and box plots display the distribution of county rates within each state.

the sorting univariate criterion, although in some examples rank 1 may in fact be the worst value, not the best.

The second issue to address is how to assign ranks when there are tied values. One convention commonly used is to assign to each of the tied values the rank equal to the average ranks of the items had they not been tied. To illustrate, suppose we wish to rank the values 5, 6, 7, 7, 7, 9, 10. The third, fourth, and fifth values are tied, so this algorithm would assign the average rank value of $4 = (3 + 4 + 5)/3$ to each of these values, so that the resulting ranks would be 1, 2, 4, 4, 4, 6, 7. The State Cancer Profiles web application uses a slightly different method, assigning the rank of the first tied value to all of the tied values, i.e., for the data above, the ranks would be 1, 2, 3, 3, 3, 6, 7. Either method should be clearly understandable to the analyst, but the NCI method does have the advantage of producing ranks that are always integers.

We explained the graphical concepts of **smoothing** in Section 3.3.3 and have mentioned the possibility of graphing model residuals. Figure 4.9 applies all of these ideas in a single graphic. The leftmost column is the same poverty data by state that we plotted in Figure 4.3. The education data in the second column of panels looks unusual—we call it a "lollipop" plot (Carr 2001). The vertical line segments connect the fitted values of this variable, similar to fitting a smooth line through a scatterplot. Lines connecting points are only shown here within each panel, but could be extended across all panels. The observed values are shown as usual by colored circles but are now connected to the smoothed line by horizontal line segments. The little white dots lighten the plot and mark the endpoints of the hidden part of the horizontal line segments.

If Figure 4.9 seems to be an unusual design, this plot can be viewed horizontally, roughly analogous to a scatterplot. The smoothing uses sorted poverty values as a predictor of college education values. The college education column, in this rotated view, can be interpreted roughly as a plot of the smoothed values versus the state poverty ranks (not the state values used in the smooth). The use of ranks distorts the view from a typical smoothed scatterplot, but the degree of distortion can be assessed by the gaps between adjacent state poverty values.

We have used an unweighted loess algorithm (Cleveland 1979) as a smoother, but many other smoothing methods are available. When the focus is on individuals rather than geographic regions, it would make sense to use population weights, but we defer discussion of population weights to Chapter 5. The District of Columbia was excluded from the smoothing because it is so unlike the other states. DC could be excluded from the plot altogether to reduce the range of the education panel, improving visual resolution, but for now we have left it as an outlier.

The objective of showing smoothed values is to provide a reference. The horizontal lines encourage discussion about which states are pulling the relationship toward more higher education (to the right) and which are pulling it down (to the left) for states that have roughly the same amount of poverty. Visually our eyes are usually drawn to the longest of these lines, which are the most discrepant from the smoothed line. This column may seem busy, but we can see a stronger negative correlation between poverty and education by looking at the smoothed education line, more so than from the original data points.

There are several design differences in the micromap panels in Figure 4.9 compared to Figure 4.8. In Figure 4.8 white space between the panel columns and around the median panels helps to convey perceptual grouping by enclosure (recall methods shown in Figure 2.10). In Figure 4.9 the white space has been dropped to increase the horizontal space available for the three wide columns. This is our preference as long as the grid line labels do not overplot and the edges of the column panels are clear.

Another design difference between Figures 4.8 and 4.9 is the boundary and background shading in the micromap panels. In Figure 4.9, the states' boundaries are light gray, the same as the background, and the states' white area shading is only slightly lighter than the background. The United States is not outlined in black. These color differences help the black outlined states to really stand out; i.e., they are clearly in the foreground. Depending on the geographic extent of the maps, deemphasizing the entire area (here, the United States) may be going too far and may lose the overall context of the maps. Drawing the U.S. outline in black for context would pop the national boundary back into the foreground, but we could introduce a conflicting foreground/background interpretation for border states that are surrounded by highlighted states, as discussed in the next section.

4.4 MICROMAP HIGHLIGHTING

You may have noticed that some states were shaded in a light yellow in several of the figures so far in this chapter. This serves to highlight states that have appeared in

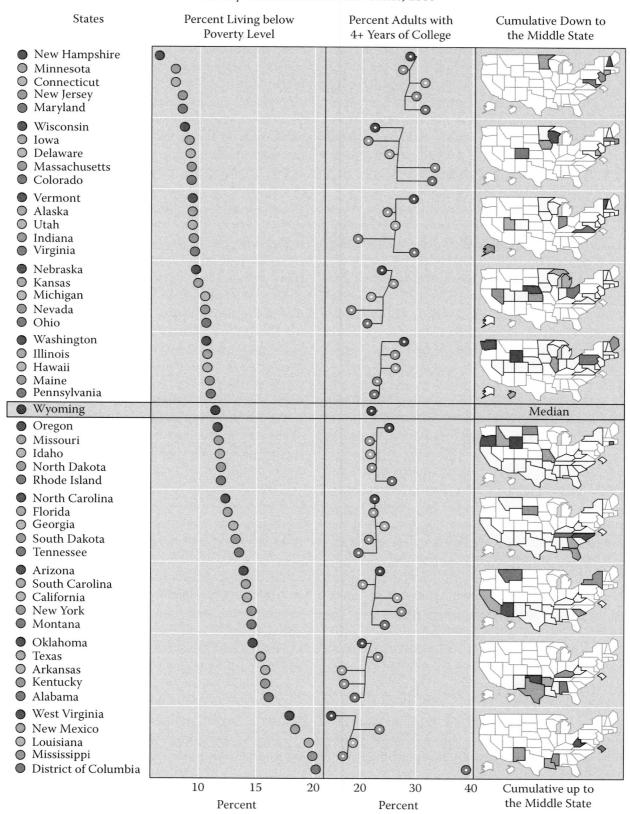

FIGURE 4.9 Poverty and education data in U.S. states, 2000, with education shown with a smoothed line (vertical) and residuals connected to the smoothes by horizontal lines.

other panels, such as those above or below each map panel. The purpose is to help the reader to see geospatial patterns that may extend across several perceptual groups. For example, the top panel of the micromap column in Figure 4.3 shows the five states with the lowest poverty rate. Four of these are in the northeast. Reading down to the next panel, we see that two other states are shaded that, when combined with the states from the panel above, form a cluster of six contiguous states along the northeastern coast. By the third panel, there are eight states in this band. The reader might miss this cluster by looking at just one map at a time.

Groups of two or more adjacent highlighted states can be of interest and trigger ideas about their similarities or interactions that relate to the sorting variable, here the percentage of households below the poverty level. We can look across a set of adjacent micromaps and mentally piece together patterns, but this is hard cognitive work and error prone. How can we do this more easily? One idea is to bring more states into the foreground than just the colored states by adding a new fill color and using black outlines on a light background.

The column of maps in Figure 4.3 (and Figure 4.5) is the simplest example. The top map panel just highlights the top five states by the usual color coding. In the second panel down, the next five states are color coded but the previous five are shown with a very light yellow fill color, and so on. That is, we accumulate yellow fill as we read top to bottom, to show states that were highlighted in one of the panels above. This reminds us of answering John Tukey's recurrent question, "How far have we come?" The foreground pattern indicates where we have focused already as we read from the top down. Now we can easily see the cluster of low-poverty states in the urban northeast portion of the United States, even though the cluster extends across three map panels. We refer to this as a cumulative fill pattern.

The goal is to select a fill color that is different than the color for background states and that does not compete strongly for attention against the colors used to link the graphs and maps for the highlighted states in each panel. The light yellow that we use here is discriminable from the background based on hue and is less saturated than the linking colors. Other combinations of line color and fill can work if they address the tasks of highlighting some states and lifting other states into the foreground. In Figure 4.3, states that are not yet shaded or highlighted remain white. We have often reversed the colors for the background states. That is, we fill the states in light gray and use white outlines (Figure 4.5).

In that design, the background states can use the same panel fill color as the color in the statistical panels.

There are many other ways to select which states to bring into the foreground of each panel in order to show patterns. An alternative method to the simple cumulative fill, used by Carr, Wallin, and Carr (2000), highlights all counties in each quartile of the sorted map values. These methods are straightforward but can appear busy when many of the regions are in the foreground, as in the simple accumulation designs shown in Figures 4.4 and 4.5. For this reason, we introduce another method that accumulates fill from both the bottom and the top of the micromap toward the middle, as seen in Figures 4.8 and 4.9. For example, use of this design in Figure 4.9 focuses attention on the half of the states with the highest (or lowest) poverty values in all of the map panels below (or above) the median bar. We can look at any of these panels, especially those on either side of the median, to judge broad geographic patterns. The map just below the median in Figure 4.9 reveals a pattern of high-poverty states covering the southern United States from coast to coast. This cluster extends northward up the West Coast (except for Washington) into the Pacific Northwest and eastward into part of the Northern Plains states. All that changes from panel to panel is the color of the states being highlighted in the sequence of perceptual groups. The highlighting by strongly saturated colors dominates the light yellow fill color.

Because Figure 4.9 is sorted by poverty from low to high values, it may seem incongruent to refer to the highlighting as "cumulative down to the middle state" when in fact poverty values are increasing in this direction. This violates the visual metaphor, where higher values are placed higher on the graph. Most readers will probably be most comfortable with a high to low sorting order, consistent with graphing conventions and the visual metaphor, while others can quickly shift perspectives and ignore violations of the visual metaphor. There will be occasional tasks that will be better served by a low to high sorting order, but this could cause some confusion when combined with highlighting above and below the median, as was done in Figure 4.9.

Before discussing the low percent poverty states, we pause to discuss the occasional difficulty that occurs as illustrated by Montana in Figure 4.8 (third panel from the bottom). Montana may appear to be in the foreground because it is outlined in black, the United States is outlined in black, and the neighboring states are foreground states and thus outlined in black. Only careful attention to the gray fill color versus the light yellow reveals that Montana doesn't belong in the foreground. A combination of background white fill, as in Figure 4.3, with a black U.S. outline, as in Figure 4.8, would exacerbate this problem. Using a foreground color that is quite different than the background white would clarify the shading of these boundary states, but as indicated above, we don't want this to compete for attention with the high-priority highlighted states, so we prefer to use a gray fill instead. It also helps to use a gray line to outline the unshaded states and the United Staets to avoid confusion with the darker outlines of background highlighted states. Monmonier's (1993) state visibility map had white space between states. This would avoid and thus solve the problem of overplotted state outlines. However, our maps are small and we strive to make the area for the color fill in small states sufficiently large for easy recognition. This example illustrates the kind of challenge posed by multiple objectives in graphics design and the use of multipurpose graphics elements. Until a better solution is found, we compromise in favor of helping the reader to perform the dominant task, i.e., spotting the highlighted states in each map panel.

Returning to the patterns in Figure 4.9, consider the map panels above the median row, which have values below the median. Now the states with the lowest poverty levels are in the foreground. Interestingly, if we look at the two panels on each side of the median bar, it may not seem visually obvious that the foreground patterns fit together to make a whole map. We don't think the problem is that we included the median state (Wyoming) in both foregrounds. Rather, we conjecture that the background pattern in one panel does not seem identical to the foreground pattern in the other panel even though the same states (minus Wyoming) are involved. Thus, it may be beneficial to bring both halves into the foreground as shown. One may be better for seeing patterns or easier to describe than the other.

The accumulation pattern in the micromaps can speed our search for a particular state. For example, if we want to find Ohio, we can glance at the map panels just above and just below the median in Figure 4.9 and note which of these has Ohio in the foreground, i.e., shaded by one of the main colors or the highlighting color. It is in the panel above the median panel, so we can scan for Ohio in the top half of micromap panels to see exactly where it is highlighted. We don't need to scan the entire micromap column, just half of it.

To summarize, we have presented methods of supplementing the color fill of the five states in each map panel with a light color fill to bring other states into the foreground for purposes of better identifying geographic patterns across the map panels. This extra fill can be applied to all states above or below the median or can be an accumulation of highlighted states from the top down, from the bottom up, or a combination of these. The designer needs to choose the method (if any) that best suits the task at hand and the expected technical skills of the audience.

4.5 MULTIVARIATE DATA

Occasionally we have more variables of interest than can easily be shown in a plot designed to display one variable per column. The multivariate aspect of the data makes it difficult to establish a consensus or at least a convention of what is "best." What is the best single estimator of air pollution when there are hundreds of hazardous air pollutants, each with their own toxicities and concentrations that vary over space? The interactions among pollutants make hazard assessment even more complex. The establishment of the U.S. Consumer Price Index is a rare example of a controversial but widely used index for multivariate data. How can we use micromaps for this type of complex data set? Let's first look at an example of bivariate data, the simplest type of multivariate data.

A standard linked micromap could easily display separate columns, but it can be difficult to identify patterns that would be evident in a scatterplot. Carr showed a micromap column of scatterplots with highlighted points for each region (Carr 2000), but overplotting and vertical resolution were problems. The graphics could instead display a density estimate of the bivariate data distribution (Scott 1992), but highlighting specific region values still poses a problem. Even more complex is the situation where each region has a large number of bivariate values. In this case, we can associate a bivariate summary with each region using a new design that incorporates the hexagon bin method illustrated in Figure 2.2.

Carr and colleagues used hexagon bin plots (Carr 1991) to summarize the nearly five hundred thousand observations of precipitation and growing degree-days across the United States, with locations classified into ecoregions (Carr et al. 1998b; Omernik 1995). Figure 4.10 defines these regions and presents the data in univariate box plot columns. As noted, this ignores correlation between the variables, so they further summarized the data into bivariate box plots, shown in Figure 4.11.

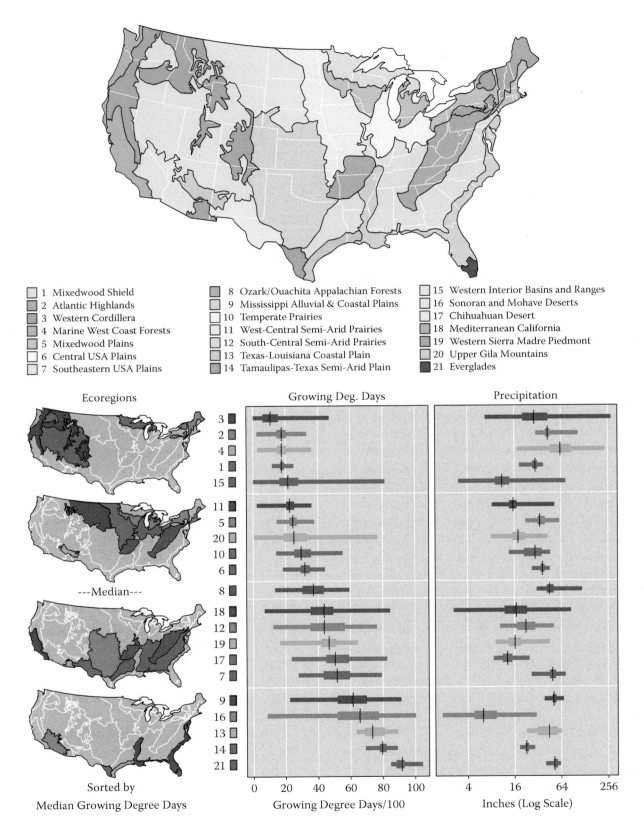

FIGURE 4.10 Linked micromap plots for the number of growing degree-days and precipitation by ecoregions for the coterminous United States (Carr et al. 1998b; Omernik 1995; Carr et al. 2000. With kind permission from Springer Science + Business Media.). The top map defines the ecoregions, with identifying numbers in the legend corresponding to row labels on the micromap plot.

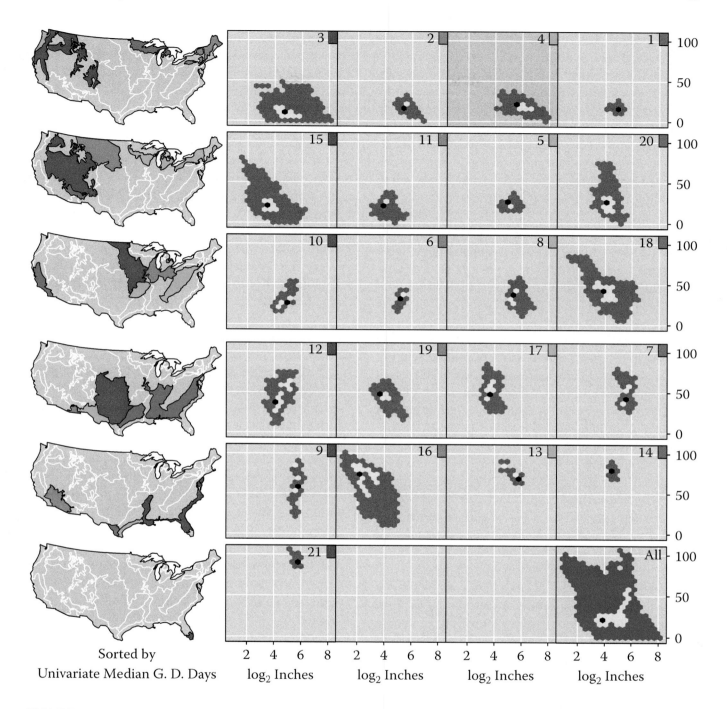

FIGURE 4.11 Linked micromap with bivariate box plots, designed to display large amounts of data in a small graph. Data are the average (log scale) precipitation (x) versus growing degree-days/100 (years) for U.S. ecoregions and for the total United States (Carr, Olsen, Pierson, and Courbois 1998b). Yellow cells are 50% high-density cells, the black cell is the bivariate median, and dark gray cells show the range of the observed data.

The hexagon cells in each panel of Figure 4.11 show the footprint of the bivariate values for each ecoregion. The yellow cells are high-count cells and contain approximately 50% of the total cell counts, akin to the content of the interquartile range in a standard univariate box plot. The black cell is a kind of bivariate median obtained by progressive erosion of counts from the yellow cells until only one cell is left. The ecoregions are ordered by median growing degree-days in Figure 4.10. In Figure 4.11 the perceptual groups contain four ecoregions each, and the color link connects each ecoregion in the perceptual group to a binned scatterplot. The ecoregion number appears by the linking tab in the top right of the scatterplot.

Several patterns emerge with this data representation. First, variation within any single ecoregion is much less than for the United States as a whole (bottom right panel), regardless of whether variation is measured by range (gray cells) or by the middle 50% (yellow cells). Few of the distributions resemble bivariate normal data. About half of the ecoregions show a negative association between precipitation and growing degree-days, as evidenced by a footprint that angles from the top left to the bottom right of the plot. Large footprints suggest partitioning the ecoregions for greater homogeneity at least in terms of the two variables. The bifurcation of yellow cells for row 5, column 2 might trigger some hypothesis generation about ecoregion 16. Does it cross a mountain range? The binned plot at the bottom right shows how different the combined scatterplot can look from its constituent scatterplots.

Many data sets that we wish to explore have more than two dimensions, so this plotting method will not work. In another ecoregion example, Carr and Olsen (1996) published a summary of 8 million pixels of land cover data classified into 159 categories. The first attempt to display these data was with a line height plot where the height of each bar represents the class acreage as a percent of each ecoregion's total acreage (Figure 4.12, left panel). That is, they created a categorical bar plot using very thin vertical bars so that all 159 bars would fit on a single page. Class labels would not be readable on the axis, but could be shown on a separate legend page or by mouseover in an interactive application.

While the amount of information compressed into this graphic is impressive, it has a high visual intimidation factor (VIF) and no patterns jump out at us. How can we simplify the look of this plot and make it more usable? One method would be to group the 159 classes into broader categories and redo the plot to show line height plots for each category in a separate column. Carr and Olsen (1996) did this using five categories—agriculture, rangeland, forest, wetlands, and tundra—and added micromaps for geographic context. As shown in Figure 4.13, it is obvious that the Everglades in Florida have more wetlands and the southwestern deserts are barren, but more detailed patterns were still difficult to see. In the next section we discuss another method that can simplify the plot.

4.6 MULTIVARIATE SORTING

Not surprisingly, the accuracy of judging similarities and differences increases with the closeness of the items to be compared (Cleveland 1985). Sorting the graphic defines perceptual groups by bringing similar items together, making them easier to compare. The advantage of sorting has been shown for dot plots (Cleveland 1985, 1993), box plots (Becker and Cleveland 1993), and tables (Wainer 1993). In addition to making the plot look simpler and less intimidating to the user, the number of visual focal points is reduced and the distance that the eye needs to travel down the page in order to make comparisons is shortened (Carr 1994). For these reasons, our convention is to sort the linked micromap plot by values in one of the columns. This puts geographic regions with the most similar values adjacent to one another. Interactive linked micromap applications allow the analyst to choose the column and direction (ascending or descending) of the sort.

How can we sort rows of multivariate data? The simplest method is to choose a single representative value, such as the median for box plots, the average value over a time series, or the sum of all types of airborne pollutants. Sometimes there is a natural or commonly used index available, such as the Consumer Price Index, which summarizes prices paid for many items in various regions of the United States. If no such index is available, we can create one by a statistical method, such as principal components analysis. Other methods have been implemented in the R package *seriation* (Hahsler, Hornik, and Buchta 2008), such as multidimensional scaling, hierarchical clustering, and the traveling salesperson problem method. Casual readers may not need to know how to implement these different algorithms but are probably familiar with the concept of the traveling salesperson problem, where an optimum ordering of the salesperson's stops is derived mathematically by using an algorithm that is much more efficient than examination of all possible pathways. Conceptually, this is the goal of all of these algorithms.

Before these algorithms had been implemented in R, Carr and Olsen (1996, 2000) used the minimal spanning tree breadth traversal algorithm (Friedman and Rafsky 1979) to sort both rows and columns of the Advanced Very High Resolution Radiometer (AVHRR) data. The result is that the rows are sorted so that adjacent rows are most similar in this multivariate sense, as shown in the right panel of Figure 4.12. Compared to the unsorted panel, more of a pattern is apparent on the sorted plot. Ecoregions 7 to 9, 13, and 21 seem similar, as do ecoregions 1, 2, 5, 6, and 10. Adding ecoregion maps makes it immediately apparent that these are contiguous areas—the former is approximately the southeastern quadrant

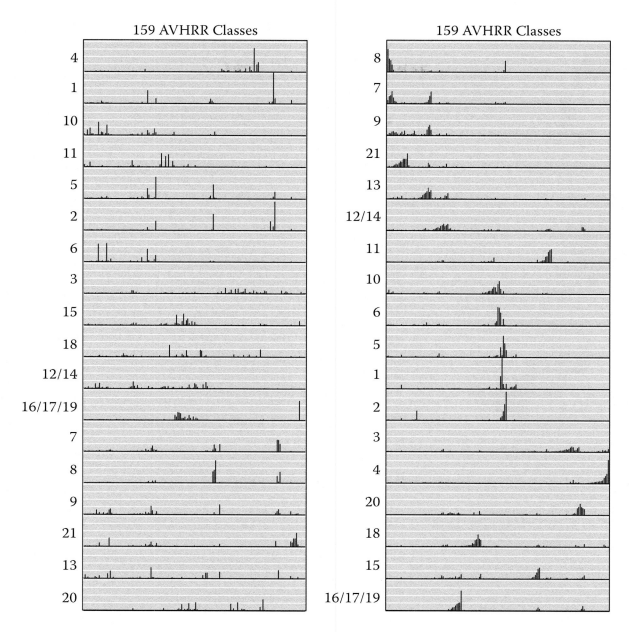

FIGURE 4.12 Line height plots of percent of ecoregion acreage for 159 land cover classes, as measured by Advanced Very High Resolution Radiometer (AVHRR) satellite imagery. Horizontal white grid lines mark 10% increments in acreage (Carr and Olsen 1996). Left image is in order of original ecoregion number*; right image has both rows and columns sorted by a minimal spanning tree breadth traversal algorithm. *Note that the ecoregions were renumbered and ecoregions 12 and 14 as well as 16, 17, and 19 were combined to match the definitions in Figure 4.10.

of the United States and the latter is New England and the Great Lakes area (Figure 4.13).

4.7 PUSHING THE ENVELOPE

So far in this chapter we have presented several variations of linked micromap plots. Figure 4.5 is the most common design. The bivariate box plot (Figure 4.11) and line height plots (Figure 4.13) are certainly pushing the envelope. The time series plot in Figure 4.14 is another

specialized micromap design. If the time series are relatively smooth, this can be an effective display method for comparing trends, but presents several design challenges. First, the choice of **sorting variable** is unclear—should the panels be sorted by the beginning value, the median value, or something else? In this example, the states are sorted by their maximum rate and displayed in descending order from top to bottom. The top-down micromap sequence shows a progressive filling in of states in the

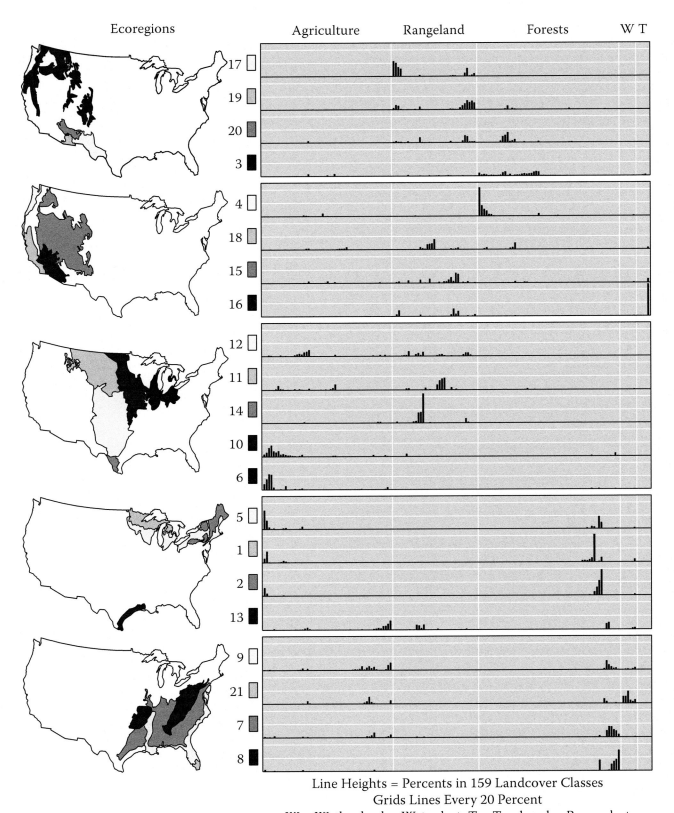

Line Heights = Percents in 159 Landcover Classes
Grids Lines Every 20 Percent
W = Wetlands plus Water last, T = Tundra plus Barren last

FIGURE 4.13 Micromaps added to line height plots of percent of ecoregion acreage for 159 land cover classes, as measured by Advanced Very High Resolution Radiometer (AVHRR) satellite imagery, grouped by and sorted within broad classes. (Carr et al. 2000. With kind permission from Springer Science + Business Media.)

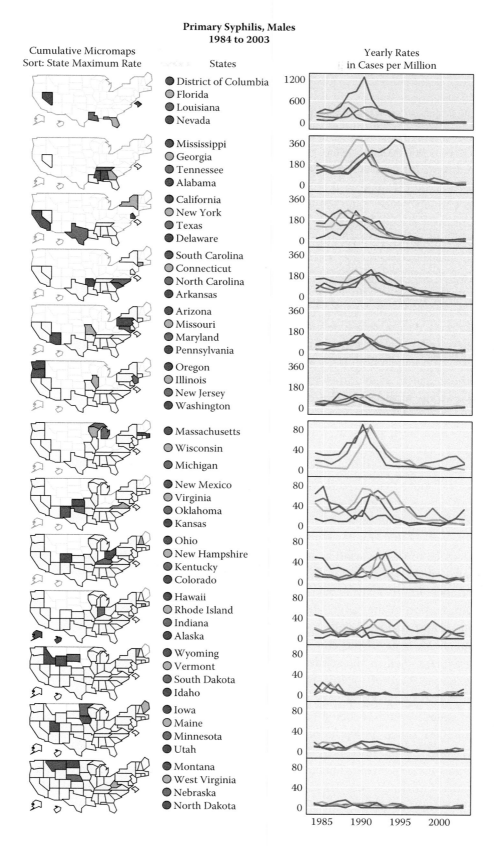

FIGURE 4.14 Linked micromap design with time series plots of primary syphilis rates in males by state, 1984–2003. (Centers for Disease Control and Prevention, National Center for HIV STD and TB Prevention (NCHSTP) and Division of STD/HIV Prevention 2005.)

south, then states along the coasts and in the Mississippi River Valley. Nevada is unusual as the only interior state in the top panel. The states in the bottom panels have had low rates over the entire period and so provide a positive story in contrast to the high-rate states.

Another challenge is **overplotting** when there are multiple time series in a panel. The lines or points may overplot in many places, making perceptual groups of five appear quite complex. Groups of three would be much clearer. However, when designing a single-page linked micromap plot for fifty-one states, fewer states per panel means more rows of panels. With more rows, the plot becomes more visually intimidating and the vertical resolution for micromap and time series panels decreases, which is not desirable. Figure 4.14 illustrates a compromise design where twelve panels show values for four states and the middle panel shows values for three states. The original design developed for CDC included five states per panel, as shown in several previous examples, and small annual points as well as the connected lines. Without careful evaluation with respect to specific tasks it may be hard to declare a winner between these two grouping choices.

Overplotting can also be a problem with labels. You may have noticed that the panel scales and grid lines were chosen so that the grid line labels don't overplot. If you have not, note that the zero value grid line is located slightly above the bottom edge of each panel. Setting the scale so the minimum and maximum rates plot closer to the panel edges makes better use of the precious vertical space to show variation across the states, but then the grid labels will overplot. The design picked nice round numbers for grid labels, but this could convey a distorted impression of the range, since some values exceed the maximum grid label value.

The range of values for times series can vary dramatically from state to state, making the **scaling** choice a challenge. In this example the 1990 syphilis rate in the District of Columbia was over 1,200 cases per million, while the majority of states never exceeded 80 cases per million from 1984 to 2003. Since the variation of individual state values is often of interest, using the same scale for all of the states is problematic. This masks the variation among states with much smaller rates. The design of Figure 4.14 compromises by using three different scales for the times series. To call attention to the different scales, the design places a gap between panels when the scale changes and uses three different panel background colors to indicate the three different scales. In this figure, horizontal grid line labels are placed on

the left in the hope that the reader will notice the scale labels when reading from the state names into the time series plot.

Even though the scales are different, some comparisons can be made across the panels. It is reasonable to scan down from the highest rate for the District of Columbia to see what other state rates also peaked around 1990, e.g., Georgia, Massachusetts, Wisconsin, Michigan, and Virginia. While their low values are less obvious because of the scale, even Alaska and Minnesota have a local high at about the same time. A number of other states peaked between 1990 and 1995, but every state had reduced rates after 1995. What might account for the reduction? Experts at CDC's Division of Sexually Transmitted Diseases Prevention can probably explain it.

Another complex linked micromap design is the display of scatterplots in column panels. It is possible to show all the points and lines in the background, to highlight subsets, and to distinguish between previously visited region values and those yet to be visited. This design has a challenge similar to the time series design. Unfortunately, sorting based on the x or y coordinate values in scatterplots or by a summary statistic such as the mean, median, or maximum value of a times series tends to produce more overplotting of highlighted values. We can borrow methods from experimental design to produce sequences of statistics that will reduce undesirable overplotting, but this can break apart patterns we were hoping to see. Usually we make our compromise toward options more likely to reveal patterns of interest.

Another variation that deserves more research is a change to an egocentric view, i.e., a presentation relative to the analyst's own perspective. We don't wish to distort the micromaps, but the statistical values can be calculated relative to the analyst's choice of reference place. This is akin to computing comparative mortality ratios, where the mortality rate for each geographic unit is divided by the U.S. reference value, except that each analyst can choose his or her reference location interactively. A value of 200% in location A would indicate that the value of the underlying statistic in place A is twice that of the selected reference location.

Another application of an egocentric view is for migration data. In Figure 4.15, we see patterns of migration of people to and from Iowa between 1995 and 2000. Color-coded lines have been added to the choropleth map linking each state back to the state of interest, Iowa. In-migration is shown in the left-hand column of panels and out-migration is shown on the right. We see

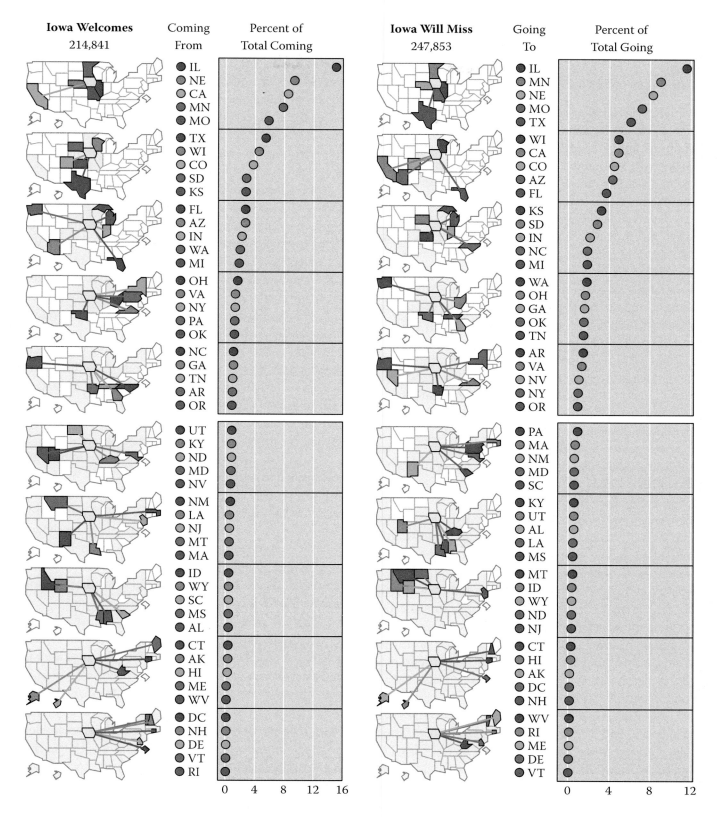

FIGURE 4.15 Showing migration using a linked micromap plot. Data are the percent of people moving to and from Iowa between 1995 and 2000.

immediately that most migration is to and from states adjacent to Iowa and to the west of Iowa.

4.8 SOFTWARE

We wrap up our tour of linked micromaps by discussing software implementations for producing them. These include PC-based and web-based programs and tools for R. We anticipate that additional software will become available from both public and private sources in the near future.

The National Cancer Institute uses linked micromaps for in-house examination of cancer incidence data and to communicate with health planners across the United States. The former system is a PC-based stand-alone program and the latter is web based, but both are interactive applications that allow the user to select data and display options tailored to their needs.

The NCI Quality Control (QC) linked micromap program was implemented using Java (Statistical Research and Applications Branch and National Cancer Institute 2009; downloadable from http://gis.cancer.gov/tools/micromaps and available on the book website). We guided the development of this software for use by NCI cancer registrars in their quality control efforts. Cancer registrars extract information from hospital records and other sources to provide data on cancer occurrence for analysis by the National Cancer Institute and the Centers for Disease Control and Prevention (CDC). The NCI Surveillance Epidemiology and End Results (SEER) registrars are required to maintain the highest standards for data quality. The QC software helps to identify and address problems locally rather than waiting for national review cycles. The registrars use this program most often without the maps to investigate variations in error or missing data rates by abstractor or hospital within their registry area. This program is very general and can read shape files and data sets in comma-delimited format.

The State Cancer Profiles web-based system was also developed at NCI in collaboration with CDC (http://statecancerprofiles.cancer.gov). This website provides numerous views of cancer-related statistics, including by linked micromaps. NCI and CDC use this usability-assessed applet to communicate cancer statistics over the web to health planners across the nation. The views are region-centric graphics; i.e., you can choose a region of interest, such as your state, and see how it compares with other states. If your state has the highest mortality rates of all the neighboring states for a particular disease,

it might motivate a region-centric look at related risk factors, or more aggressive action to prevent, diagnose, or treat the disease. The current implementation offers data on cancer rates, demographics, and selected risk factors, all selectable from drop-down lists on the left side of the screen (Figure 4.16). A referent or target line can be specified by the user, such as the green target area in Figure 4.16. Mousing over one of the linked elements causes all to blink.

The interactivity allows people to choose from among a variety of options and supports geographic focus on U.S. states or counties of a selected state. Viewing options include state- or county-centric views that compare a selected state or county to other areas at the same geographic scale. Clickable icons control the sorting of panels. Variables available for selection include cancer incidence and mortality rates, demographic variables, and risk factors. A continuing drawback is the limited screen resolution and space consumed by agency branding and other required text. The 8.5 × 11-inch page can show all the U.S. states, while the web version requires scrolling. In addition, there is currently a substantial difference in resolution between a printed plot and a web-based graphic.

Symanzik modified the Java code from NCI to implement a similar system for the display of West Nile Virus cases at Utah State University (http://webcat.gis.usu.edu:8080/index.html; Symanzik and Carr 2008). Changes in rates or case counts over time can be shown by two color-coded dots connected by a thin line in a single panel, similar in purpose to the arrow plot in Figure 4.8. The color of the line indicates direction of change: red for increasing, green for decreasing.

Another web-based implementation of linked micromaps is the display of acreage, production, and yield of harvested cropland by the National Agricultural Statistics Service at the U.S. Department of Agriculture. This website (http://www.nass.usda.gov/research/sumpant.htm) first presents a summary row plot for states, then presents a univariate linked micromap plot on demand for any of the row plot columns. Unlike the NCI web application, these plots are static, so the user cannot choose content to display other than the precalculated plots (Symanzik and Carr 2008). A more interactive linked micromap web application is the display of species diversity developed by the EPA's Regional Vulnerability Assessment Program (http://waratah.com/willametteedt/welcome.asp). Although limited in geographic scope to the Willamette River Basin, it does

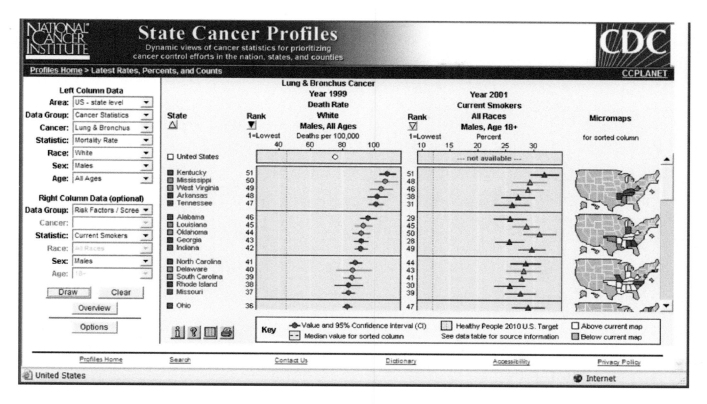

FIGURE 4.16   Screen shot of linked micromaps output from the State Cancer Profiles website (http://statecancerprofiles.cancer.gov) showing the data selection pick lists on the left. Note that the scrolling occurs within the page, i.e., the labels remain in view as the micromaps scroll.

allow the user to select variables to display in either a single map or two maps with an explicit difference map.

4.9   SUMMARY

In this chapter we have presented basic features of linked micromaps, some design considerations, and designs for more complex data. The linked micromap design is the most mature type of micromap. It has been used in various forms for over a decade and has proven to be an extremely useful graphical method for exploration, analysis, and presentation. After extensive usability studies, only minor changes were made to the original design for the NCI web application. This is an indication that reliance on the principles summarized in Chapters 2 and 3 will lead to a highly intuitive and usable design. As the use of linked micromaps expands, more and more types of data will be explored in this way. As more software becomes available, we expect this to become a popular data visualization method.

5 Conditioned Micromaps

5.1 INTRODUCTION

Our micromap tour now continues with a detailed look at conditioned micromaps. The goal of this type of micromap plot is to help us think about the geographic patterns and associations in our data using more than one variable at a time. Conditioning partitions the geographic regions shown in a single choropleth map. When conditioning is based on the values of two other variables, the resulting two-way layout of micromaps highlights different subsets of the regions. This conditioned structure often reveals patterns that beg for explanation, encourages hypothesis generation and further study, which leads to better understanding about the relationships among three variables. Put simply, most people recognize and have ideas about map patterns even if they do not yet understand all of the details about the processes that generated the data being mapped. The widespread appeal of maps and easy interaction with the dynamic sliders of conditioned micromaps encourage users to become more involved, learning more about the data collection process and even about the statistical methods involved. Thus, conditioned micromaps provide a wonderful framework for data exploration that incorporates both pattern discovery and hypothesis generation.

The development of the conditioned micromap was motivated by the need to explore patterns in the newly published National Center for Health Statistics (NCHS) *Atlas of United States Mortality* (Pickle et al. 1996) in 1996. Rates for many of the causes of death had not been mapped before, and analysts began speculating about the causes of the notable patterns. There was a need for convenient, interactive software that would allow exploration of associations between the mortality rates and hypothesized risk factors, prior to investing scarce resources to fully model each cause of death. Carr combined the ideas of Cleveland's (1993) conditioned graphs and Tufte's (1983) use of small multiples of juxtaposed panels into a display of small multiples of juxtaposed maps that highlight subsets of regions, conditioned on two other variables (Carr, Wallin, and Carr 2000). This provided a way to display and interact

with three variables for regions on a map. This approach was motivated by John Tukey's (1979) advice to map fully adjusted variables. Conditioning does not provide fully adjusted variables but is a step in the right direction toward controlling sources of variability that can mask patterns. Whether Tukey's target of full adjustment can be achieved depends on having suitable data for the task.

The goal of showing two variables using a choropleth map was not new. Two early papers from the statistics community considered the use of color and shrinking the area of each region in proportion to the mapped value, as illustrated in Figure 2.2 (center) for hexagons (Carr, 1980), and bivariate color maps (Wainer and Francolini 1980). Drawbacks were noted for both approaches.

Several related designs were proposed in 1992 publications: Monmonier's (1992) principle of a "cross map" was independently implemented by Carr, Olsen, and White (1992), and Cleveland, Grosse, and Shu (1992) published two-way conditioned, juxtaposed scatterplots and explanatory margin panels that showed the levels of conditioning. A prototype HealthVisB system proposed by MacEachren et al. (1998) used a bivariate color scheme to represent mortality rates (by hue) and levels of a risk factor (by light and dark shades). This system provided animation of the maps over time but was limited to the display of a health outcome and a single, dichotomized risk factor.

The conditioned choropleth maps shown in Carr, Wallin, and Carr (2000) included the dynamic conditioning slider concept but generated the graphics using Splus® (Insightful Corp. 1988) script files. The dynamic three-class masking slider in Explor4 (Carr and Nicholson 1988) and the sliders described by Shneiderman (1992) left no doubt about the feasibility of dynamic response. The sliders in Carr's subsequent Java implementation (CCmaps) provided dynamic response for both maps and statistics (Carr, White, and MacEachren 2005).

A single choropleth map is the beginning stage of conditioned choropleth maps. Choropleth maps typically show just one variable, which we will call the study

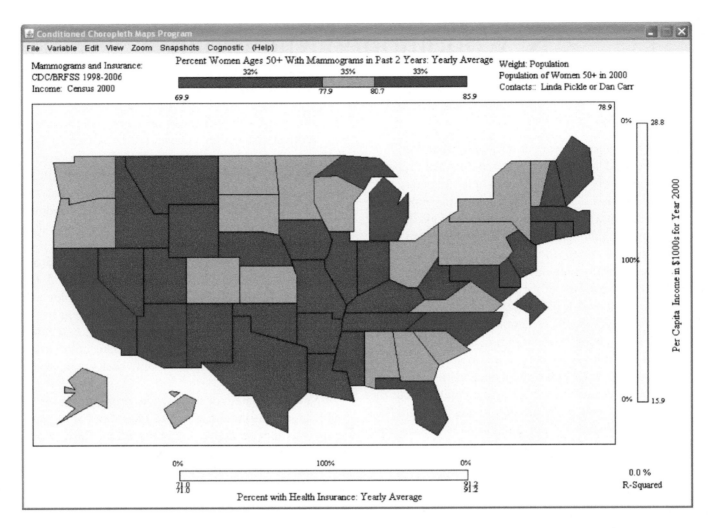

FIGURE 5.1 Map of the percent women ages 50+ who received a mammogram in the past two years (yearly average over 1998–2006). **Note that red is used for low (bad) values, contrary to traditional use.**

variable. The study variable label may tell us what and when, and the colored regions may tell us how much and where, but without additional variables we cannot begin to answer a host of questions. Why this pattern? How does this pattern relate to what I already know? We understand the world in a multivariate space-time context. Our graphic design needs to help us to think visually about the relationships among multiple variables while retaining the geospatial context of the study variable. This is not an easy task. However, conditioning the single map by two variables will allow us to make some progress within a framework that is cognitively tractable. We can study patterns within a modest number of subsets of regions and make comparisons across a small number of well-organized panels that highlight these subsets.

We introduced conditioned micromaps in Section 1.4. To review, we reproduce Figures 1.6 and 1.7 with additional annotation here. One generalization that we can make from the full map of mammography rates in Figure 5.1 is that a higher proportion of women are screened in coastal states. (Note that we have reversed the usual color scheme of red for high and blue for low values in order to highlight regions in need of intervention by the red color, representing danger or alarm.) A cancer control planner might wonder whether women who do not receive mammograms according to the medical guidelines do so because they have limited access to mammography facilities (less likely now that screening facilities are widespread) or because they cannot afford to pay for the service. To explore the latter hypothesis, we partitioned the states on the mammography rate map according to per capita income and percent of residents

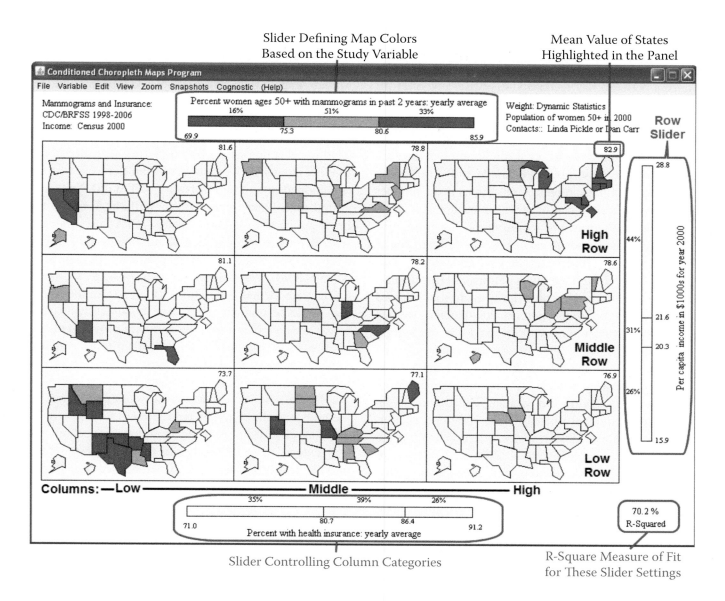

FIGURE 5.2 Annotation defining components of conditioned micromaps, using CCmaps software for illustration. Maps are of the annual percent of women ages 50+ who received a mammogram in the past two years, averaged over the period 1998 to 2006, with states stratified by their residents' per capita income and health insurance coverage.

with health insurance, as shown in Figure 5.2. We can see that most low screening states (shaded red) fall into the bottom left panel (low income and low health insurance coverage), suggesting that low screening rates are associated with a lower ability to pay, because of either lack of insurance coverage or insufficient income to self-pay.

Before we get too far along in our discussion, it is important to understand what conditioned micromaps are and what they are not. Conditioned micromaps define subsets of geographic regions based on a cross-categorization using the conditioning variables that describe the regions. Combining the highlighted regions of the nine micromaps

in Figure 5.2 (in set theory terminology, a union of maps) exactly reproduces the map in Figure 5.1. Each state appears in one and only one of the nine panels. The panel in which a state appears depends on the state's average per capita income and percent of residents with health insurance coverage. Conditioned micromaps do *not* display different study variables in the different micromaps. For example, when we have two study variables such as male mortality rates and female mortality rates, we can produce two complete micromaps indexed by gender. We will address the study of one- and two-way indexed study variables in the next chapter on comparative micromaps.

Epidemiologists are familiar with indexing disease rates by a hypothesized risk factor that is categorized into three or more classes. Evidence of a dose-response relationship between risk factor and disease is one of Hill's (1965) criteria for arguing that an association is causal. However, what epidemiologists call indexing is usually a conditioning of individuals within each geographic area by the stratifying variable, not a conditioning of the geographic areas themselves. For example, in Figure 5.2 the denominator of the screening proportions is the number of women ages fifty and over within each state. These maps could be redrawn to represent the screening proportions among women in each state who fall into one of three categories of income (three rows) and who do or do not have health insurance (two columns). In this display, the bottom left panel would have each state shaded according to the proportion of low-income women without health insurance who received a mammogram in the past two years. That is, the denominator of the screening proportion in that panel would be the number of low-income women without health insurance in each state, not the total number of women. This display might be of interest in answering different questions, such as the epidemiologist's dose-response question, but it is not a conditioned micromap plot. Conditioned micromaps categorize geographic regions, not individuals within those regions. Users of conditioned micromaps need to be aware of this important difference, as geographic patterns of individual characteristics are not necessarily well represented by maps of aggregated data (the "ecologic fallacy"; Robinson 1950).

This chapter is organized somewhat differently than Chapter 4. There we focused primarily on the micromap design elements and the example data, leaving discussion of available software to the end. The design principles for conditioned micromaps follow those for linked micromaps, with the addition of interactive slider bars and several statistical tools. In order to illustrate these interactive features, discussion of the CCmaps software, included on the website for this book, is integrated with the explanation of the design and implementation of the methods. Dan's early conditioned micromap designs (Carr, Wallin, and Carr 2000) were criticized as impractical for use with complex maps, as for the 3,100 U.S. counties. He and his students implemented the designs using Java in order to demonstrate rapid response to the slider movement for large databases, something not possible using other platforms at that time (Carr et al. 2002).

Now, with faster computers and more available GUI tools, implementation on other platforms is possible,

as long as functions are available for mapping and for two-threshold sliders. It should be feasible to produce versions in a scripting language such as R, Splus, Matlab, and Python that have reasonably fast updates and that provide ready access to broad functionality. For example, Michael Friendly (2006) has written a SAS® (SAS Institute, Inc.) macro that displays a static conditioned micromap given user-specified thresholds. In 2003 Jung Jing Lee and Kyu Won Lee sent us their conditioned micromap software (STAGIS) that supported more than three classes. William Smith at the Environmental Protection Agency (EPA) has produced several versions of conditioned micromaps, including a dynamic web-based system (http://www.turboperl.com/dcmaps.html). We will use our own software, CCmaps, to illustrate the method in this chapter so that the reader can replicate our examples and dynamically explore his or her own data using the software provided on the book website. We discuss the statistical tools available in CCmaps in some detail, as they are not necessarily included in other implementations.

In the next section we will discuss categorization methods and the rationale behind them for the simpler one-way classification. Then in following sections we will extend these principles to two-way designs and provide details of visualization methods for conditioned micromaps and examples of their use.

5.2 ONE-WAY CONDITIONED LAYOUTS

5.2.1 CONTINUOUS VERSUS CATEGORICAL MAPPING

After all of our discussion on accurate representations in previous chapters, you may be wondering about the poor variable resolution created by converting each of the continuous study and conditioning variables into three ordered classes. This can result in a tremendous loss of detail. In CCmaps, our conditioned choropleth software, this loss is partly ameliorated by the dynamic conditioning sliders that allow us to focus on an arbitrarily small interval of values. Later in this chapter we will describe statistical measures that are more precise and easier to communicate than visual impressions of color-coded maps. These measures can also serve as live statistical feedback to help reduce the information loss by a judicious choice of conditioning category cutpoints.

An argument against presenting visualizations of the original continuous variables is the cognitive burden they impose on the reader. For example, the same mammography data shown in Figure 5.1 are presented in Figure 5.3 with a nearly continuous color scheme

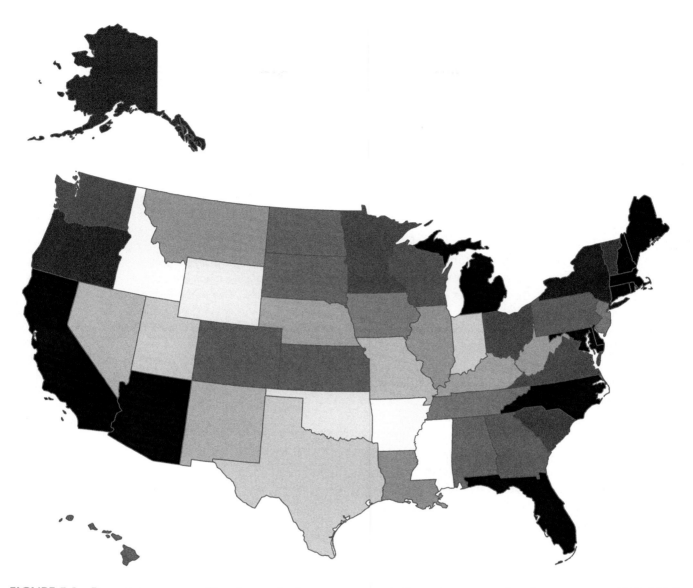

FIGURE 5.3 Percent women ages 50+ who received a mammogram in the past two years (yearly average over 1998–2006) shown with a continuous color scheme (lowest to highest values represented by light to dark gray). Compare to patterns in Figure 5.1.

(the map has forty-five distinct gray shades due to ties among the states). We can see that Massachusetts and Delaware have the highest screening rates, a fact that was masked by the categorization in Figure 5.1, but the colors of some other states are so similar that we are unable to rank them. Is Iowa higher or lower than Montana? This type of color scheme is difficult to use under ideal conditions because of the subtle differences among the colors. Limitations of printing or computer display resolution make things worse.

Another difference between Figures 5.1 and 5.3 is that CCmaps uses map boundaries that are smoothed caricatures of the true boundaries, a slight adaptation of Monmonier's (1993) visibility base map. This map has

simplified the boundaries of each state and has enlarged the area of the smallest states while retaining the relative position of the forty-eight contiguous states. We do not need the detail of Figure 5.3, e.g., jagged coastline boundaries, to discern patterns in the data. Further, placement of small polygons representing Alaska and Hawaii in the lower left corner of the map layout (Figure 5.1) results in a more compact layout than in Figure 5.3, a real advantage when displaying many small micromaps on one page.

5.2.2 One-Way Conditioned Scatterplots

Scatterplots display the patterns of the values for two variables. Figure 5.4a shows a scatterplot that introduces a new data set, male homicide rates for U.S. states in 2001–2005. The scatterplot shows a strong positive, nearly linear, correlation with the percent of households headed by females. We will soon look further at this relationship and at other variables.

In anticipation of the conditioned micromaps to come, Figure 5.4 shows the colors that will be used to distinguish the low, middle, and high values of the homicide rates. The blue, gray, and red are a diverging three-class color scheme suggested by ColorBrewer (www.colorbrewer.org) that can be used by color-blind readers. Both sequential and diverging color schemes (see Figure 2.7) have merit when the class levels are ordered. The sequential scheme is natural, following the visual metaphor of light to dark representing increasing class order. However, the diverging scheme used in this example draws attention to the two extreme classes better by giving the middle class less visual contrast from the background. Choosing between the two types of color schemes is not an easy call.

You might be wondering why we chose three categories instead of some other number. We need at least three categories in order to examine any pattern for a nonlinear trend, i.e., to detect curvature in the association between two variables. Three is small enough that the conditioned micromap display can show the 3×3 grid of maps; imagine how small the micromaps would need to be if we wanted to display a 5×5 grid on a single screen! Three is also cognitively usable—we can easily name and remember the categories as low, middle, high, or small, moderate, large.

Figure 5.4b is a conditioned scatterplot of homicide rates versus the percent of female head of household that highlights subsets using methods similar to those of the maps in Figure 5.2. In an analysis by Cubbin, Pickle, and Fingerhut (2000, 90), this variable was of interest because social scientists hypothesized that patterns of homicide rates in a recent atlas (Pickle et al. 1996) could be related to social problems such as poor housing. A strong effect of this variable is also consistent with social disorganization theory that posits that lower levels of family stability lead to less effective social controls against violence. The left panel of Figure 5.4b highlights states with low percent female head of household. The values for the other states are marked with small gray circles to reduce overplotting while maintaining the context of the highlighted points within the full data set. Similarly, the middle and right panels highlight states with middle and high percents of female head of household, respectively. Note that

A Spatial Analysis of Homicide Rates

Cubbin, Pickle, and Fingerhut (2000) found that homicide rates in health service areas (HSAs) during 1988–1992 were significantly higher in the most urban areas, but sociostructural differences also were important. These factors appeared to operate similarly for both black and white males. Of the sociostructural factors measured, income inequality within HSAs was much more important than median income level for whites, but both income level and inequality were significant for blacks. This is consistent with social deprivation theory and previous findings of a threshold effect of income on homicide; i.e., above some absolute level of income, higher rates are seen in areas with greater disparities of income regardless of the income level, whereas in lower-income areas both are important. The proportion of households headed by women also affected homicide rates, consistent with social disorganization theory that argues that lower levels of family stability lead to less effective social controls against violence. Of course, caution against overinterpretation is in order—death certificates report only the address of residence of the victim, not the location of the homicide, but most homicides occur close to home.

the left panel is more homogeneous in terms of colors; i.e., most of these points are blue.

Statistical descriptions can help support (or refute) our visual impressions from both scatterplots and maps. The mean is a common statistic used to characterize the center of a set of values. The horizontal lines in Figure 5.4b show the mean for the filled dots in each panel. The means are clearly increasing as we look from left to right. Those with statistical eyes will also note that the vertical variability about the mean is also increasing as we look from the left panel to the right panel. There are many possible statistical descriptions of these data, such as a simple linear regression with just two parameters, an intercept and slope. However, we will keep things simple and just summarize using means.

5.2.3 One-Way Conditioned Micromaps

When we look at the one-way conditioned maps in Figure 5.5 the geospatial structure related to percent female head of household is pretty remarkable. The left panel has a strong latitude component; i.e., most of these are northern states. Maybe there is a relationship to population density. A wild conjecture is that men and women marry and stay married to keep warm in the winter. Although these ideas are speculative, even

Male Homicide Rates, 2001–2005

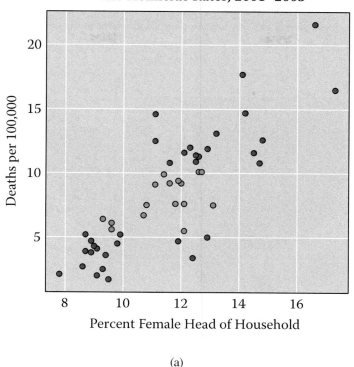

(a)

Male Homicide Rates 2001–2005

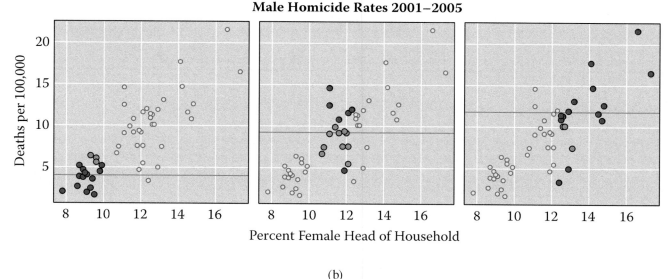

(b)

FIGURE 5.4 (a) A scatterplot of age-adjusted male homicide rates by state (excluding DC) for the years 2001 through 2005. The three colors indicate low-, middle-, and high-rate states with seventeen, sixteen, and seventeen states in the three classes, respectively. (b) Conditioned scatterplot of male homicide rates by state (excluding DC) for 2001–2005. The left panel highlights states with low percent female head of households. The middle and right panels highlight states with middle and high percents of female head of household. Horizontal lines are panel means.

silly, you can see how these map patterns can motivate hypothesis generation.

Surprisingly, the geographic patterns by state are similar to those in maps at the health service area (HSA) level (aggregations of counties; Makuc et al. 1991) over a decade earlier, even though the homicide rates have been declining overall (Pickle et al. 1996). We present state data for simplicity of visualization, but social factors can vary greatly across small neighborhoods, so an analysis at a finer geographic resolution would be preferred.

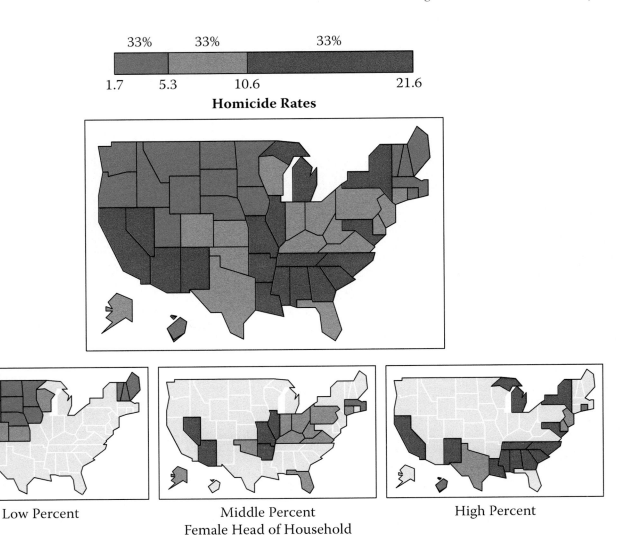

FIGURE 5.5 Mapping homicide data from Figure 5.4. Top: Male homicide rates by state, 2001–2005, with blue, gray, and red representing lowest, middle, and highest rates, respectively. Bottom: Same homicide rates conditioned on percent female head of household, with left to right micromap panels representing lowest, middle, and highest values (with approximately 33% of states in each category).

As we saw for scatterplots, the mean can summarize the level of the rates on a map. This simple description of the regions on a map by their average value is very difficult to compute by looking at the colors of the states, even when the legend is available that defines the color classes. Our eyes respond to the areas of regions, so rough visual impressions can be quite poor because large areas are often sparsely populated. For example, tiny Delaware had more residents in 2000 than the larger states of Alaska or Wyoming. In addition, mentally computing an average of values on a map is cognitively very difficult.

How can we represent the mean of a map? Map axes are geospatial coordinates, so the lines that we used in Figure 5.4b won't work for representing the means. For conditioned micromaps we show mean values in the corners of the separate panels. Other statistics might be of interest to display, such as the standard deviation or coefficient of variation of the mapped values. This sequence of maps could include scatterplot smoothes, which we will discuss in a later section.

5.2.4 An Introduction to Sliders

Three-class sliders are a key element of dynamic interaction for conditioned micromaps. The basic purpose of the slider is to assign the values of a continuous or ordered variable to one of three classes that we describe as having low, middle, and high values. The sliders in CCmaps have two mouse-adjustable thresholds that

TABLE 5.1
Example of Slider Threshold Settings to Define Three or Fewer Categories

Threshold Position		Number of	
Lower	Upper	Categories Displayed	Example Slider
Inner	Inner	3	
Minimum	Inner	2	
Inner	Same as lower	2	
Maximum	Maximum	1	
Minimum	Maximum	1	
Minimum	Minimum	1	

bound the middle class. The slider algorithm assigns values between the two bounds to the middle class, values below the lower bound to the low-value class, and values above the upper bound to the high-value class. The implementation assigns values tied at the upper bound of a class to that class, not to the next higher class. By this rule, if the lowest boundary of the slider was set at the minimum data value, that value would not be included in the lowest class. Consequently, the slider's fixed minimum boundary is made slightly less than the minimum data value so that all data values will be included on the slider bars.

This slider implementation allows the easy creation of maps with fewer than three classes and corresponding class colors on the map. Sample slider settings are shown in Table 5.1 with white dots added to clarify the location of the lower and upper adjustable slider thresholds. For example, moving the lower threshold all the way to the left edge of the slider (lower dot moved to the left) displays only the middle and high categories (row 2 of table). Moving the two thresholds so that they coincide displays only the low and high categories (row 3 of table). Both of these settings emulate the operation of a two-class slider. Likewise, the thresholds can be positioned so that only a single class is displayed.

Figure 5.6 provides an example of filtering using a conditioning slider. Here the top study variable slider has both thresholds moved to the left so that all states fall into the high-value class, represented by red. Likewise, the right conditioning slider has a light blue fill, indicating that all states fall into one class. Consequently, states are not categorized by homicide rates (top slider) or by population density (right slider). The bottom slider is set at the median value for percent female head of household; states with values equal to or less than the median are shaded light yellow, which indicates the filtered out interval on the slider and the filtered out regions on the map. States above the median are unshaded, so that the default red shading from the top slider appears. The result of these three slider settings is not a map of homicide rates but a binary map indicating where the state percent female head of household is high or low. This is an example of using CCmaps as a filter for one of the slider variables.

The CCmaps sliders illustrate design choices made to address usability issues. The first decision was to make the slider scale linear in terms of its unit of measure. Since the original unit of measure, shown below the bar, is often chosen to be something subject matter experts will understand, a linear scale provides a more useful context than a slider based on the percent of regions included. Of course, the percent of regions in each slider interval might be important, and is shown above the sliders, but the slider does not change linearly by these

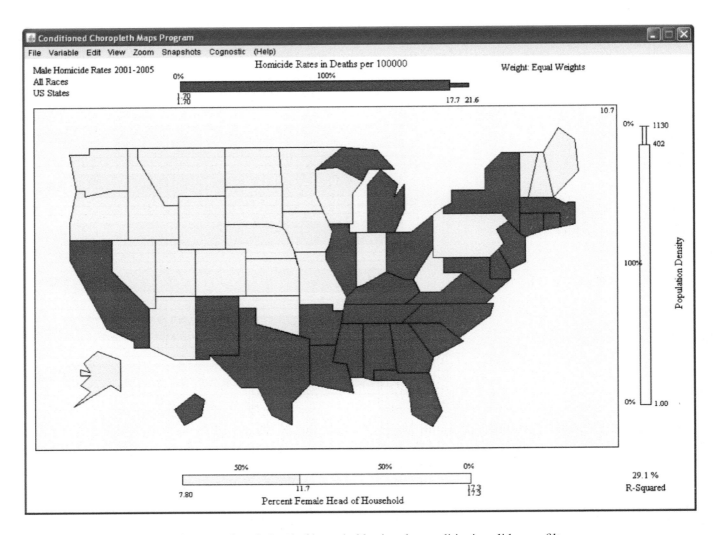

FIGURE 5.6   Binary map of percent female head of household using the conditioning sliders as filters.

Homicide Rates in Deaths per 100,000

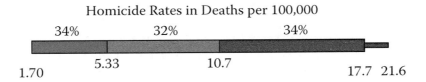

FIGURE 5.7   Example slider bar for mapped study variable.

percents. Figure 5.7 shows an example slider for homicide rates in deaths per one hundred thousand population. The values below the slider at its left and right ends show the minimum and maximum values for the states, respectively. The two values below the center portion of the bar, 5.33 and 10.7, label the two interior thresholds bounding the middle class. The number of significant digits displayed is based on interpolation between the endpoints and should not be used to infer that the data values are accurate to that degree. The vertical distance between the slider threshold labels and the slider endpoint labels avoids label overplotting. However, some overplotting does occur in the current implementation when the two thresholds are placed very close together.

It is usually easier to remember percents with nice rounded values, and so we often start by adjusting the sliders to approximately 20% each for the low and high classes, leaving 60% in the middle class. This design draws more attention to the two extreme classes by color, compared to the gray shading for the middle class. Relegating the middle class to a visual layer between background and foreground by using a less saturated gray shading can make the graphic appear less complex. With only 68% of the regions highlighted by a strong foreground color (reds and blues in Figure 5.7), we can quickly shift attention from one class to the other, especially when strong patterns are present. Other strategies can also guide our choice of percents. For example, we may want to limit our extreme classes to only the top and bottom 5% of regions. If we have no other preferences, we could choose to split on quartiles (25%, 50%, 25%) or tertiles (33% each) until we have a better feel for the data and could partition into more homogeneous groups.

As we become familiar with a particular type of data, our growing experience and knowledge become increasingly tied to the units of data. We gain a sense of what differences or ratios are of practical importance and what would be considered just random variation that we aren't going to worry about at the moment. In this case, we will probably prefer to choose the slider cut-points based on the actual values rather than percents. However, we will probably still be inclined to pick nice values for slider thresholds because this is easier to remember and to communicate.

Data transformations are sometimes helpful in terms of providing a more uniform resolution between values across the range of the slider and in terms of studying the association with other variables. Some software, such as Gapminder World (Gapminder Foundation 2009), has a logarithmic transform option. This is useful when variables have positive values and a long right tail. When the values are all greater than 1, the log transformation reduces the gaps between the large values. Cleveland (1985) recommends the use of log base 2 when the range of data is modest. When the range of values is large, more people may be familiar with a log base 10 scale, i.e., displaying the power of ten.

The connections between gaps in the data and slider resolution may be thought of in the following way. Large gaps between values at the extremes of the distribution consume a relatively large fraction of the total slider length. Slider-defined thresholds are discrete values based on pixel size and the number of pixels available from the minimum to maximum values (currently 440 in CCmaps). In the high-density part of the distribution, usually the middle, the number of data values to the right of the slider threshold can jump substantially even when the slider is moved right by only 1 pixel. In other words, the slider provides poor resolution in the high-density part of the data and unnecessarily high resolution in the very low-density part of the data. The log transformation provides a better threshold resolution balance for linearly scaled data when the data distribution is very

skewed. The cost is that the units of measure become less familiar to many people.

CCmaps uses a piecewise linear slider to deal with extreme slider resolution imbalance caused by outliers, providing better resolution in the main body of the data, i.e., in the middle of the distribution. This construction is signaled by the appearance of a thinner bar on one or both ends of the slider. The thin tail has a limited fixed length, bounded by the extreme value and the adjacent value (see Figure 4.7), on the given side of the distribution. Two criteria trigger this change. First, there is at least one outlier as defined by the box plot criterion (see Figure 4.7). Second, changing the scale from the adjacent value to the extreme value on that side of the distribution needs to cause the slider values to change values per pixel noticeably faster. On occasion the gap between the extreme and adjacent values is small, so there is little advantage to changing scale. To illustrate, compare the length of the red portion of the slider bar in Figure 5.7 from the threshold of 10.7 to the end of the thick portion of the bar, 17.7, with the length of the thin portion representing values from 17.7 to 21.6. In the presence of extreme outliers, dragging the mouse over the thin bar can cause the threshold value to change very quickly. The merit of piecewise linear slider bars will become clearer in Chapter 7, when we encounter large population changes before and after the 2005 hurricanes.

5.3 TWO-WAY LAYOUTS

5.3.1 INTRODUCTION TO TWO-WAY LAYOUTS USING TABLES

The basic structure of the conditioned micromap can be illustrated by a table where the rows and columns are defined by two variables (Figure 5.8). Each cell of the table, which will become a single panel in the conditioned micromap graphic, will contain a single map with regions shaded only if they fall into the categorization for that cell, e.g., high values for both variable 1 and variable 2 in the upper right cell. The shaded regions on each map are color coded according to the values of the study variable.

5.3.2 CONDITIONING IN TWO-WAY LAYOUTS

The conditioning that we use highlights each state in exactly one micromap panel in a grid of micromap panels, with the panel choice determined by the row and column conditions (as in Figure 5.8). The states not meeting the slider conditions are put in the background

Low 1, High 2	Mid 1, High 2	High 1, High 2	High	
Low 1, Mid 2	Mid 1, Mid 2	High 1, Mid 2	Middle	Variable 2
Low 1, Low 2	Mid 1, Low 2	High 1, Low 2	Low	

    Low   Middle   High

Variable 1

FIGURE 5.8   Structure of a conditioned micromap according to values of two conditioning variables.

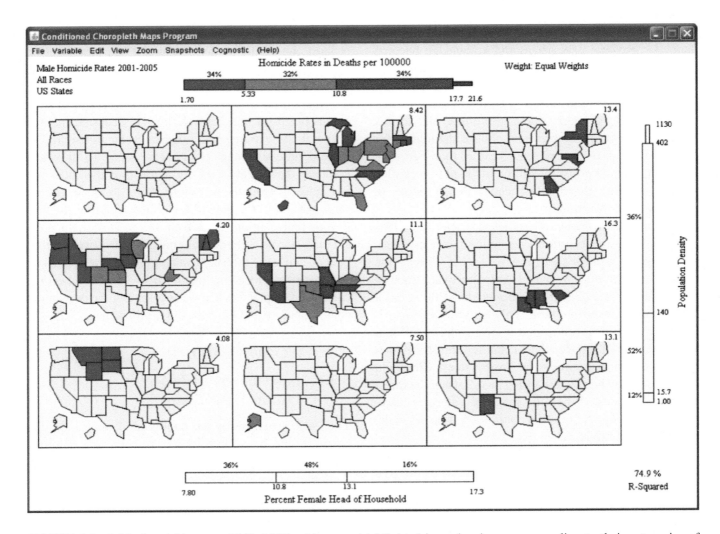

FIGURE 5.9   Male homicide rates, 2001–2005, with states highlighted in each micromap according to their categories of percent female head of household and population density.

by using a common background color. We prefer maps that appear light rather than maps that appear dark, so we have chosen light yellow as the background color. You may have different color preferences.

The conditioning used in this book always partitions the regions into subsets. Each region with study variable values is highlighted in exactly one of the nine micromaps. Becker and Cleveland (1996) described a more general form of conditioning that can produce

overlapping subsets called shingles. While shingles have merit in the context of smoothing, we keep it simple and partition the regions into disjoint subsets.

In Figure 5.9, the male homicide rates have been categorized by population density as well as percent female head of household. The content of each cell in this two-way layout is determined by the slider settings and the row and column of the cell (compare to Figure 5.8).

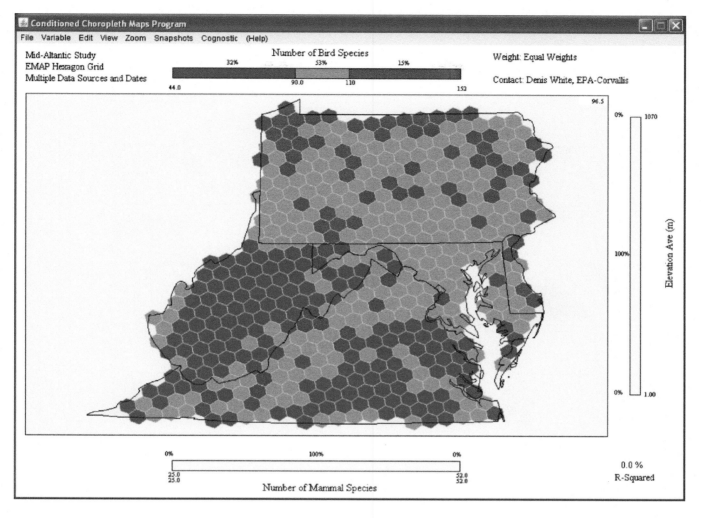

FIGURE 5.10 Use of a hexagon grid in conditioned micromaps to display the number of bird species in local regions. The blue hexagons have 90 or fewer bird species, the gray regions have 91 to 110 species, and the red hexagons have over 110 species (up to 152).

The next example illustrates how conditioned micromaps can be applied to very large datasets by using hexagon binning, defined in Figure 2.2. As seen in the linked micromap ecoregions of Chapter 4, the mapped regions do not need to be administrative areas such as states and counties. Environmental monitoring research has made tremendous methodological advances in the past 15 years, e.g., in how to sample in space and time in order to provide valid assessments. Researchers in EPA's Environmental Monitoring and Assessment Program (EMAP) used hexagon grids to develop local summary statistics that could be used as the basis for an environmental analysis. White et al. (1999) describe multiple sources of data and the target application of studying biodiversity. Figure 5.10 displays the number of bird species found in each hexagon cell whose centroid falls in a Mid-Atlantic state. The preponderance of lower count cells are in West Virginia.

Figure 5.11 conditions these bird species counts on the number of mammal species and the average elevation in the hexagon area. These were chosen for illustration from sixty variables available in the variable menu. The slider settings are near optimum for describing the number of bird species by these two conditioning variables, chosen using an algorithm we will explain in Section 5.5.1. Looking across the middle row, we see that there were fewer birds in regions with fewer mammals (blue cells on the left) and more birds where there were more mammals (red cells on the right), suggesting a positive correlation between the numbers of species of birds and mammals. The average number of bird species per cell increases from left to right in this row—81.8, 97.1, 108. The bottom row, lower elevation areas, does not follow this pattern, but we note that higher numbers of bird species are seen along the shorelines in the left and middle panels, in contrast to the inland cells highlighted

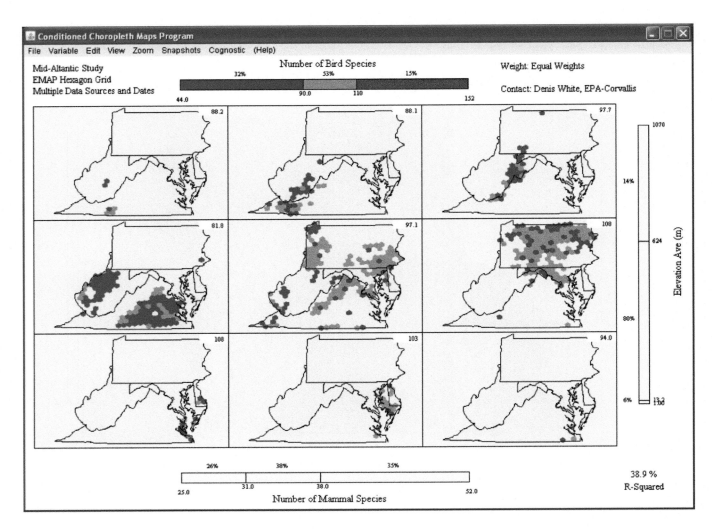

FIGURE 5.11 Hexagon bin conditioned choropleth micromap of bird species data (Figure 5.10) conditioned on the number of mammal species and average elevation.

in the rightmost panel. We could search through the remaining fifty-seven variables on the data set seeking stronger associations. This leads us to our next discussion about conditioning variable selection, but we note here that White and Sifneos (2002) illustrated the use of regression tree methodology in a study of similar data for the state of Oregon.

5.3.3 Choices for Conditioning Variables

So far we have described the conditioning sliders and have illustrated them by conditioning variables hypothesized to be predictors or correlates of the study variables (mammogram use, homicide, or bird species). What other variables might we choose? In addition to other risk factors that might serve as predictors, we might wish to condition on a confounding variable, i.e., one that alters the association between another predictor and the study variable and is itself associated with both of these

variables. For example, the association between heavy coffee drinking and the incidence of low birth weight is confounded by smoking. Women who smoke are more likely to be heavy coffee drinkers and also to have low birth weight babies. The excess risk of low birth weight due to heavy coffee drinking is modified from 85% to 53% by controlling for smoking (Van den Berg 1977). Since a major purpose of conditioning is to control the variation in the study variable, this helps us see any remaining spatial variation that may be related to other variables, i.e., other than the conditioning variables. For example, an application of CCmaps to low birth weight data could look for geographic patterns across levels of coffee drinking (controlled by one slider) while conditioning on the level of cigarette smoking (controlled by the other slider). The reader is cautioned to compare appropriately adjusted rates in CCmaps. In this example, comparison of rates unadjusted by age could

State Age-Adjusted Rate

	Above United States	Similar to United States	Below United States
Rising Significantly	Arkansas Indiana Kansas Kentucky Louisiana Maine Ohio Oklahoma Tennessee	Alabama Georgia Iowa Michigan Mississippi Montana Nebraska North Carolina Pennsylvania South Carolina South Dakota Wisconsin	Colorado Hawaii Idaho Minnesota North Dakota Utah
Trend of Rates Stable	Delaware Missouri Nevada Oregon West Virginia	Alaska Connecticut Florida Illinois Maryland Massachusetts New Hampshire New Jersey Rhode Island Vermont Virginia Wyoming	Arizona New Mexico
Falling Significantly	None	District of Columbia New York Texas Washington	California

FIGURE 5.12   Presentation of state age-adjusted death rate/trend comparisons for lung cancer, all races, and females only, death years for most recent significant trend period through 2005. (Redrawn query results from NCI's State Cancer Profiles website: statecancerprofiles.cancer.gov)

be misleading because it would be unclear how much of the regional differences in rates was due to differences in the number of older women at higher risk of having low birth weight babies or to differences in the proportion of smokers in each state.

If the study variable is available for multiple time periods, it can be helpful to condition simultaneously on the level and time trend of the study variable. NCI's State Cancer Profiles website does this for age-adjusted mortality rates. We display a cross-tabulation of state lung cancer mortality results for females in Figure 5.12. Following the NCI format, the level of the state rates compared to the U.S. rate is stratified by the columns and the time trends are stratified by the rows. More precisely, the conditioning is based on the results of statistical significance tests of these comparisons. For example,

Delaware's rates are significantly above the U.S. rate, and so Delaware appears in the left column. Delaware appears in the middle row that is labeled "stable" because its trend is not significantly different from a flat trend (slope of 0). Although the NCI website displays this type of information in tabular form, as shown here, each cell of the table could display a map of rates in the most recent time period, as we have done in Figure 5.13. The layout is like a conditioned choropleth representation with sliders for two three-category conditioning variables. This text layout that has been highly successful on the State Cancer Profiles website was motivated by a 3 × 3 layout of conditioned maps. The maps add value to this table by showing, for example, that most of the states with high rates and an increasing time trend (upper left cell) form a contiguous band of states in the central United States.

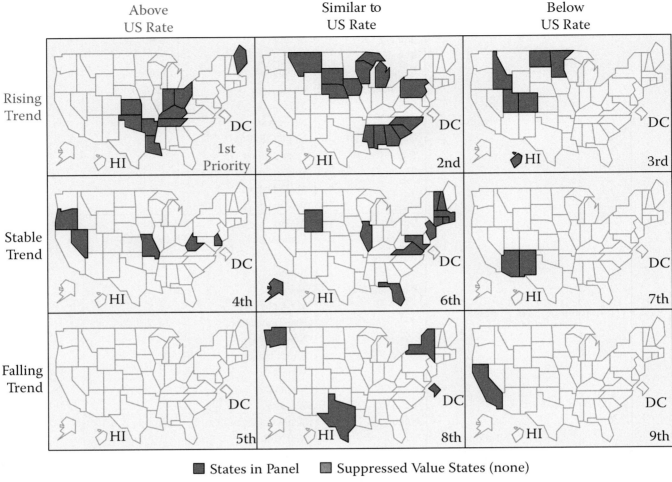

FIGURE 5.13 State age-adjusted death rate/trend comparisons for lung cancer, all races, females only, death years for most recent significant trend period through 2005, redrawn from Figure 5.12 with map display.

Another conditioning variable of interest is one measuring the reliability or quality of the data. The statistical significance used in Figure 5.12 is one measure of reliability, but others are the relative standard error or coefficient of variation or simply the number of events from which the study variable was calculated. Conditioning the number of events as above or below some threshold of reliability will separate the "noisy" regions whose rates are based on small numbers from those with more reliable rates. A more qualitative but related conditioning variable is the source of the data, if there are multiple sources of varying quality.

5.4 HIGHER-ORDER LAYOUTS

Theoretically we are not constrained to two-way layouts, but the added complexity of higher-order layouts

imposes practical limitations. For example, just adding one more conditioning variable with only three categories would result in twenty-seven (3 × 3 × 3) micromaps! As the number of conditioning variables or categories increases, the size of each micromap is smaller and fewer and fewer regions are shaded on any one map. Not only is it more difficult to see patterns on each very small map, but the cognitive burden of interpreting, remembering, and comparing the many micromaps also increases. Multivariate smoothing could be used to stabilize the patterns in conditioned maps, as indicated in Section 3.3.3 (Whittaker and Scott 1994). Becker and Cleveland (1996) proposed a rectangular array format for the display of higher-order plots called trellis plots because of their resemblance to a garden trellis pattern. The conditioning variables are each categorized into intervals, often overlapping, and subsets of the data meeting the conditioning values are plotted in each panel graphic.

5.5 DESCRIBING AND COMPARING GROUPS OF REGIONS

5.5.1 COMPARISON OF MICROMAP MEANS

Once we define different groups of regions to highlight in different micromaps, a helpful next step is to describe the study variable values for the highlighted regions in each micromap. As we saw for one-way layouts (Figure 5.4b), the mean provides a brief objective description for thinking about and comparing groups of regions. The visual impressions based on the highlighted region colors often draw our attention to interesting patterns. However, more processing is necessary to quantitatively characterize the patterns for more precise description, comparative reasoning, and communication.

For example, we saw in Figure 5.5 that homicide rates were relatively low in northwestern states. In Figure 5.9 the full map is conditioned on population density (right slider) and percent female head of household (bottom slider). We might quickly notice that nearly all of the low-homicide states, highlighted in blue, are in the leftmost column of micromaps. We can characterize these panels quantitatively by their means—4.08 and 4.20 deaths per 100,000 in the lowest and middle categories of population density, respectively, the lowest values of all nine panels. This process of going from a quick visual impression to a further processed result is roughly analogous to our preattentive process of aligning and focusing our eyes on a spot in an image so we can use the high-resolution part of our eyes to figure out what it was that drew our attention. In the micromap context we can let the computational processing augment and extend our quantitative thinking about the structure that drew our attention.

We could provide a more detailed micromap description such as "the micromap in the top row and middle column has a fairly even mixture of low-, middle-, and high-homicide-rate states." This statement describes variability in the state homicide rates. The quantitative analogue would be to compute the variance or standard deviation of the highlighted values. The standard deviation of the top center micromap states is 4.76 deaths per 100,000 population, while it is only 1.94 deaths per 100,000 population for the bottom left micromap states. (Note: This is a simple unweighted calculation of standard deviations of the mapped rates

for illustration, ignoring the different population sizes in each state.) We prefer to use the standard deviation rather than the variance in this context to characterize the deviations about the mean, because it is in the same units as the data. Note that we could not compute the standard deviation for the bottom center micromap since the calculation requires two values and only one state is highlighted. We do not place high confidence in standard deviations based on just a few values. Remapping the homicide rates at the county level would provide many values per micromap, but the trade-off would be many counties with no observed homicides, i.e., a sparse data problem. A compromise is to use an aggregated county unit, such as the health service areas used in the NCHS atlas maps (Makuc et al. 1991; Pickle et al. 1996).

We are mainly focused on the means in this homicide example. We consider a simplified case of conditioning using one variable to illustrate how this focus can mask interesting variability patterns in the data. Suppose two-class conditioning separates study variable values, perhaps model residuals, into two classes as indicated by the black and red values below. Regions with the black and red values would be highlighted in separate micromaps whose rounded panel means would be 0.00 and 0.01. This suggests the distributions are very similar in terms of their means. However, an untold part of the story is that their standard deviations of 2.33 and 0.49 are quite different. The color variation in micromap panels sometimes warns when the panel means are telling only a part of the story.

$$-2.53 \;\; -1.34 \;\; -0.96 \;\; -0.85 \;\; -0.49 \;\; -0.33 \;\; -0.10 \;\; 0.24 \;\; 0.73 \;\; 2.10 \;\; 3.58$$

	Mean	SD
Black	0.00	2.33
Red	0.01	0.49

Visual impressions may be rough but can guide us to use statistics and other visual analytics that are appropriate for the data. CCmaps does not currently include the option to display the standard deviation along with the mean. This can be viewed as an advantage, as it limits the distracters when there are clear differences in the micromap means. However, it would be good to be able to turn on the display of the standard deviation when motivated by visual impressions and to have an algorithm that would detect corresponding patterns and turn on the display based on a tunable parameter. Certainly it is important to look at additional facets of the data and not just the means. In the next section we examine some statistical methods that can help us do this.

5.5.2 Data Models, Fitting Means, and the Search for Good Slider Settings

Data models provide estimated or predicted values that can be compared to the observed data values. These predicted values can be produced from simple models, such as analysis of variance (ANOVA) or linear regression, where computer calculations are almost instant, or from very complex ones, such as hierarchical spatiotemporal models that require hours (or days) of computer calculation. In general we hope that the modeled values provide a better description of some underlying phenomena than the data values themselves, smoothing away some of the background noise in the data to reveal underlying patterns. In some situations we hope that the models are useful for interpolation at other locations in the domain of interest or for extrapolating to new circumstances or future time periods. In general, we seek models that are useful for thinking about underlying phenomena and that fit the observed data well without being overly complex.

We use an R^2 statistic to compare simple models in this chapter. When studying states partitioned into a two-way grid of panels, we define

$$\text{Grand Mean} = \frac{\sum_{i=1}^{n} \text{State Value}_i}{n}$$

$$R^2 = 100 * \frac{\sum_{i=1}^{n} \left(\text{Model Value}_i - \text{Grand Mean} \right)^2}{\sum_{i=1}^{n} \left(\text{State Value}_i - \text{Grand Mean} \right)^2}$$

where n is the number of states included, *State Value*$_i$ is the observed value for state i, and *Model Value*$_i$ is the corresponding estimate or prediction resulting from the data model. We use the term *grid model* to describe a simple two-way model where the model value for a state is the mean of the panel in which it is highlighted. The means can be considered parameter estimates for this simple model. If all grid cells are occupied, there are nine parameters. Two simpler models are row and column models, where the model value for each state is the row mean or column mean of the row or column where that state is highlighted. These models usually have three parameters, one for each row or column. Another commonly used model is one that assumes independent row and column effects for a two-way table. Some models are better suited to the structure of the data than others. Sometimes a linear regression with just two parameters,

a constant and a slope, is a better model than even the simple ones described here. The choice of an appropriate model is beyond the scope of this book, but the point is that often we can see more with the help of models than we can by only looking at colors on a map.

Since our model and model parameter estimates are based on partitioning regions, such as states in our example, we can imagine using more variables to partition states until each state is in a group by itself. This perfect fit would result in an R^2 value of 100%. This is usually considered to be overfitting. That is, the model fits not only the underlying structure in the data but also the impact of mistakes, idiosyncrasies, and miscellaneous sources of variance on the observed data. Just as many early users of neural net models were proud of their overfits, it can be tempting to be proud of the fit of the simple two-condition models described here, but they can overfit the data. The assessment of models by cross-validation or by applying the simple model parameter to new data often provides a more realistic assessment of model quality. Note that we are not doing inferential statistics here, but are using this convenient structure to provide an organized and useful way to think about the two-way layout. Our purpose is illustrative, but with some refinement these methods could be extended to draw inferences about the data.

To illustrate, the grand mean of homicide rates for all states in Figure 5.9 is 9.76, and the total sum of squares for the grand mean model is 976. The grid model values for the states in the lower left panel (Montana, Wyoming, North Dakota, and South Dakota) would all be 4.08, the mean rate for that panel. Repeating this process in each panel, then subtracting the grand mean from the grid model value for each state, squaring and summing these deviations, results in a total of 731, the numerator of the R^2 calculation. The ratio of 731 to the grand mean model sum of squares (976), expressed as a percent, is 74.9%, the R^2 statistic. The interpretation in this context is that the conditioned micromap means (the grid cell model) account for roughly 75% of the discrepancy (variability) between the exact fit model (equivalent to the observed values) and the grand mean model. This value appears as feedback in the lower right corner of the CCmaps plot in Figure 5.9.

Searching for good fits is often *ad hoc*. Often a good starting point is to have about the same number of states (roughly 33%) in each slider interval for both conditioning sliders. After a while, the manual search can become tedious. In the current version of CCmaps,

440 pixels is used to display each slider. Each of the two thresholds of a slider can take on 440 values, and the lone constraint is that the lower threshold cannot be greater than the upper threshold. This means there are over 97,000 (441 choose 2) possible settings per slider and over 9.4 billion possible settings for the two conditioning sliders. We could use a little help to find the best of these many settings. In CCmaps we use a cognostic to help us find good slider settings. A cognostic was defined by John Tukey as a diagnostic to be interpreted by a computer (Tukey and Tukey 1981). Our interpretation based on usage is that a cognostic is a computation used to help humans prioritize their visualization and modeling activities. This idea was extended by Wilkinson et al. to implement "scagnostics," multiple computations typically applied to all the plots of a large scatterplot matrix in order to prioritize them for analyst review (Tukey and Tukey 1985; Wilkinson, Anand, and Grossman 2005).

The cognostic included in CCmaps is a grid search for high R^2 slider settings. This coarse grid search avoids redundant calculations and so is very fast. The algorithm returns a list of top ten settings that can be selected from a menu. Selecting an item immediately applies the slider settings to the map (or other views). On occasion there are quite different settings that will produce almost the same value of R^2. The R^2 value obtained in this way may not be the global optimum, i.e., the highest value obtainable, because not all of the 9.4 billion settings are tried. The grid search generates candidate thresholds for each slider from which it tries all possible pairs. The candidates are selected to have roughly the same number of states (combined state weights) between thresholds. A consequence of the coarseness of the grid search is that often a slight change in one or more of the thresholds can produce higher values. It can be fun to try to do a bit better than the computer. Some might prefer another algorithm, such as a more detailed initial grid search, a second local grid search using the neighbors of the slider settings from the first grid search, or a search based on the distinct values of the data.

The search runs the risk of overfitting the data with classes of two-way mean models that may not be ideally suited to the underlying structure of the data. Very simple regression models may fit the data better and hold up better when subjected to the test of new data. CCmaps is intended primarily for hypothesis generation, and it should be understood that the mean models are not

expected to be the most accurate or defendable portrayal of a conjectured underlying model. That is, the mean models are not to be taken too seriously, but they are often reasonable starting points for data exploration.

5.5.3 ANOVA-Like Descriptions

We want to analyze the variation in region values. Analysis of variance (ANOVA) is one of many statistical procedures for studying variance. We want to follow the thought process of analysis of variance even though our exploratory conditioning violates the conditions for rigorous statistical inference. ANOVA-like descriptions are still useful in an exploratory setting, but we do need to keep in mind the importance of external validation of the patterns that we find by using this method.

We can do more with the means that appear in the top right of the conditioned map panels. Figure 5.14 shows an alternative view of the nine micromap means, which is available via the ANOVA choice under the Views menu in CCmaps. The top panel (Table 5.1) shows the 3×3 layout of panel means plus the row and column marginal means and the grand unconditioned mean. Fairly often only one of the two conditioning variables is very interesting. We may learn more by looking at the margins showing the means based on this one conditioning variable. To illustrate, the column margins at the bottom of the top panel in Figure 5.14 correspond to low, middle, and high categories of the percent female head of household, the bottom slider under the maps. The column means of 4.2, 9.3, and 14.8 deaths per 100,000 population are the means of state values for states highlighted in the respective columns. The differences between these values are substantial, suggesting that conditioning on percent female head of household makes quite a difference in the homicide rates. The row marginal means reflect conditioning just on population density using the conditioning slider on the right. The smaller gaps between these means (6.1, 8.2, and 9.2 deaths per 100,000 population) suggest that the conditioning on percent female head of household controls more of the variability than does conditioning on population density.

The second table in Figure 5.14 shows main effects and interactions based on table 1. The main effects are the deviations calculated by subtracting the grand mean from the marginal means. The values in table 2 are calculated from the following equation: observed cell mean = grand mean + row effect + column effect + interaction.

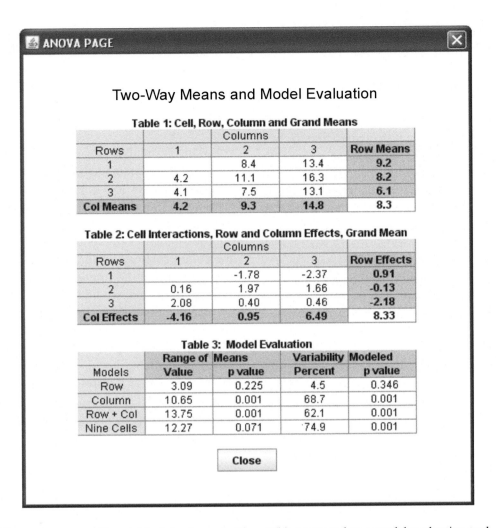

FIGURE 5.14   Two-way means of homicide data shown in Figure 5.9, presented as a model evaluation tool.

For example, the cell interaction for row 1, column 2 is calculated by solving this equation for the interaction: 8.4 (observed) − 0.91 (row) − 0.95 (column) − 8.3 (grand mean) = −1.78. We are most interested in how much the main and interaction effects vary about zero in terms of practical importance. We may not care about cell interactions less than 0.5 in table 2, but there seems to be substantial interaction in several other cells.

If the data resulted from an experiment involving nine experimental conditions in a 3 × 3 factorial experiment, it would be appropriate to consider a two-way analysis of variance. However, we have partitioned data after obtaining it in order to create the table. We have sought conditioning thresholds to bring out differences in the study variable means shown in the cells of the table. Finally, we have ignored the spatial autocorrelation in the data. This is not an appropriate circumstance for using an analysis of variance and claiming that we have statistical evidence that the patterns are real, as noted in the previous section. Our efforts are better

spent in looking for independent research, scientific knowledge about the phenomena under consideration, and additional data for analysis that many help disprove or confirm the pattern that has motivated our interest.

A further problem is that the conditioning may result in empty cells or other forms of serious imbalance in cell counts (e.g., different numbers of states highlighted in each map). There are at least two lines of statistical thought about how the analysis of variance should be performed for such unbalanced situations, which we leave as a topic addressed in more advanced texts. However, even though the ANOVA-like calculations described here are not statistically rigorous, the R^2 measure can be helpful in guiding our slider setting selection. Randomization tests, which we discuss in the next section, are less dependent on classical distribution theory and can be suggestive of whether the association observed between the study and conditioning variables is random or not. Although this is a less specific hypothesis than classic ANOVA tests, it does not violate the distributional assumptions.

An Experiment in Hypothesis Generation

When we reason about patterns it is hard to be self-aware of our predisposition to seek supporting or confirming evidence. It takes discipline to think hard about what evidence is available that might destroy our hypotheses. We have tried a simple example suggested by Wason (1960) in the classroom and have informally corroborated the result. The teacher writes down three numbers with a suggestive pattern appropriate for the audience, for example, the sequence 2, 4, 6. The teacher asks students to write down these numbers, guess a next correct number in the sequence, and state how confident they are in knowing the rule for generating the correct number. The teacher walks around to confirm the correct guesses. The process is repeated with most students guessing 8, 10, 12 and their confidence about knowing the rule increasing with each successful guess. When asked what they think the rule is, they all are confident that it is to add 2 to the previous number. The teacher tells them that they are wrong—the rule is simply an increasing sequence of numbers. After guessing 8, a guess of 8.5 would have been confirmed as correct and would have contradicted the students' guess of what the rule was, but no one really challenged what they thought they knew by guessing something inconsistent with their hypothesized rule. The moral of this story is that unless the hypothesis is tested, we can be fooled by our observations. This is true for any statistical result and data visualization, including micromaps.

5.5.4 RANDOMIZATION TESTS TO ASSESS PATTERNS INDUCED BY CONDITIONING

We look for patterns using the power of our visual cognitive processing, our computational search algorithms, and the guidance of our mental models that suggest expected patterns. Pattern description can range from rough verbal description to the calculation of a host of specific statistics. Using a quantitative description, such as means, opens the door to quantitative models of the patterns. Because our visual system is so good at detecting patterns, we need tools to help us judge whether a perceived pattern is likely a real one or is just due to chance variation.

Just as there are many approaches to pattern description, there are many approaches to pattern evaluation. For this goal, our task shifts from search, discovery, and hypothesis generation to critical thinking. Both strategic and tactical elements can be included in our searching process. We can strategically select data sets to merge and select variables to examine based on our knowledge of phenomena, data availability, and data quality. We can tactically choose to use fast algorithms and automated processes for pattern evaluation. We will describe one method for pattern evaluation by using the homicide data.

One model that looked promising for the homicide data in Figure 5.9 was conditioning the micromaps into three groups using percent of female head of household, with thresholds of 10.8 and 13.1% for defining the low-, middle-, and high-value groups. This column model reduces the variation about the grand mean by 68.7%. This amount of reduction is very helpful for describing what we see in Figure 5.9, but has our manual search for optimal conditioning slider settings resulted in a happenstance, noise-dominated pattern, or have we found something interesting? A randomization test can provide a primitive answer.

If percent female head of household is not really associated with homicide, then randomly permuting the state labels associated with percent female head of households should not affect the optimum slider choice by very much. For each permutation, we can search for new slider settings that give a high value of R^2. Doing this two-step permutation and search process 9,999 times would yield a total of 10,000 R^2 values (the original plus the permutation results). The fraction of these values that would result in a reduction of variation about the grand mean that is equal to or greater than our observed value (68.7%) can be interpreted as a p value of statistical significance. A small p value suggests that our observed association between the state mortality rates and percent of female head of household is very unlikely to be a random event. A large p value warns us not to get excited about a pattern, as it could be just one interesting pattern that we found by searching through noise.

CCmaps has not fully implemented the randomization test, which can be very slow when implemented rigorously. It keeps the thresholds from the first model, rather than optimizing the slider settings for each permutation, and it only does 999 permutations by default.

5.6 WEIGHTED DESCRIPTIONS AND COMPARISONS

The U.S. Constitution declares that all people are created equal, and created the Senate, where each state has equal representation, and the House of Representatives, where the number of state representatives is based on the state population. These unweighted and weighted approaches, respectively, to congressional representation carry over into visual representation and visual analytics. Equal representation for regions makes for a simpler analysis. There is no need to obtain weights for regions to use in the analysis, just as there is no need to conduct a decennial census to allocate votes in the Senate. However, when we are more interested in people than in the geographic units, e.g., states, clear reasoning often favors using population weights. CCmaps provides support for weighted analysis.

When using weights, it is common to scale them so that they sum to 1. Consider an example of crude death rates for states:

$$R_i = D_i / P_i$$

where D_i is the number of deaths in state i and P_i is the population.

Let the scaled population weight be

$$w_i = \frac{P_i}{\sum P_i}$$

Then the total weighted U.S. rate is

$$R_{US} = \sum w_i R_i = \sum D_i / \sum P_i$$

A weighted average for a subset of states, e.g., for a region of the United States, would be calculated similarly. It is instructive to note that in this simple example of U.S. crude rates, the weighted crude mean summed over all states is exactly equal to the crude rate for the total United States. Of course, if the mapped rates were age adjusted or computed in some other way, their weighted average may not equal the U.S. rate, but they are often similar unless the distribution of state rates is very skewed, especially if extreme rates are heavily weighted. Similarly, weighted means calculated for each panel are often reasonable (when weights are scaled to sum to 1 for each panel) and likely a much better estimate of the rate for the composite of states highlighted in that panel than an unweighted average. For example, the crude U.S. lung cancer mortality rate for white women during 2001–2005 was 50.4 per 100,000 population, very similar to the unweighted state mean rate of 50.3. However, the unweighted mean for states in the Pacific Census Division (Alaska, Hawaii, California, Oregon, Washington) during the same period was 44.7 per 100,000 population compared to the population weighted mean of 42.7. California's population is 75% of the total in this region, and so its lower rate (39.5) pulls down the regional mean.

Except for the simple crude rate example, the population weighting method implemented in CCmaps is not equivalent to using weights proportional to the inverse of the variance of the rates, a common practice in statistics to account for variability in the reliability of the state rates. In many instances, the study variable is an age-adjusted rate, which has a more complex variance representation, but the simple weighting method usually provides a reasonably accurate set of spatial patterns in the micromaps. However, the user can provide inverse variance or other appropriate weights and specify these as the weighting factor in CCmaps.

5.7 ALTERNATIVE VIEWS

The conditioned micromaps concept was developed specifically to display data on choropleth maps. However, other views can provide further insight into the patterns of the map and of the data distribution. In this section we present alternative views that have been implemented in the CCmaps software.

5.7.1 MODEL-BASED VIEWS

Two-way conditioning is a very simple modeling method used to bring out patterns. We chose 3×3 two-way conditioning as a cognitively tractable sweet spot within an unlimited number of modeling approaches. As we saw in Section 5.5.3, CCmaps can display an ANOVA-like table of row, column, and interaction effects. CCmaps does not currently include any modeling capability beyond this simple model of the 3×3 data display, but the results of more complex models can be imported into CCmaps for further examination and display.

Statistical models typically decompose observed values into fit and residuals. Mapping fitted values shows broad patterns that may help us to understand and explain the process that generated the data. Mapping residuals can show us a mixture of noise and anomalies. Sometimes we are more interested in the broad patterns, but at other times we wish to identify the anomalies, e.g., where some corrective action needs to be taken.

There are many different kinds of models that lead to fits and residuals. Models attempt to fit local data and can differ substantially in variables and scales used. Simple geospatial smoothing of point data could be thought of as a simple model—it involves just one variable and a function that summarizes values at neighboring locations, with neighborhood defined by either pairwise distances or adjacency of locations. Figure 5.15 shows the results of a more complex hierarchical spatial

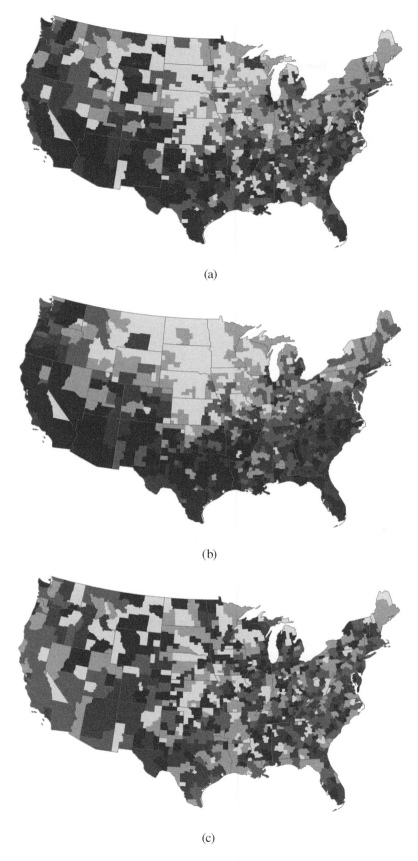

(a)

(b)

(c)

FIGURE 5.15 Mapping model results. (a) Observed average annual homicide rates for white males ages 15 to 24, 1988–1992, color coded according to quintiles of the distribution (darker is higher), (b) model predicted rates, and (c) residuals from model (b).

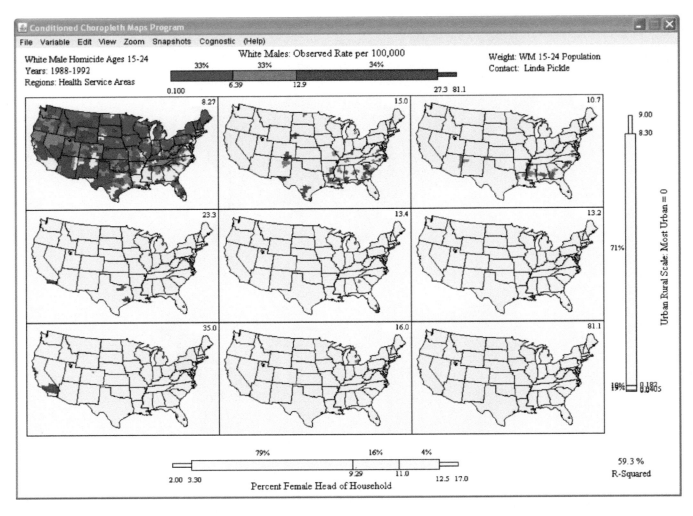

FIGURE 5.16 Conditioned micromaps for observed white male homicide rates, 1988–1992, by health service area, conditioned on urban/rural status and percent female head of household.

random effects model. The observed data underlying white male homicide rates for 1988–1992 (ages 15 to 24 shown in Figure 5.15a) were modeled by health service area—805 geographic units that are aggregations of counties (Makuc et al. 1991). Use of this data set follows our earlier examination of homicide data at the state level but provides more observations on smaller geographic units to demonstrate model-based views. The Cubbin model that included socioeconomic and other regional covariates found that urban/rural status (Butler and Beale 1994), income inequality, and percent of households headed by women were among the strongest predictors of homicide (Cubbin, Pickle, and Fingerhut 2000). Rates predicted by this model for young white males are mapped in Figure 5.15b. We still see the strong pattern of high rates in the South and West that were apparent in the original map, but there is less visual "noise" in the predicted map, as we would expect from a well-fitting model. The map of residuals from this model

(Figure 5.15c) has no obvious spatial trend or clustering, another sign that the model fit the data well.

Figure 5.16 shows these same homicide data in a conditioned micromap view, conditioned on urban/rural status (0 = totally urban, 9 = totally rural) and percent female head of household, with the slider settings determined by the CCmaps cognostic. The R^2 is 59.3%, surprisingly good considering the complexity of the model needed by Cubbin, Pickle, and Fingerhut (2000) to fit this data set. (Remember that our R^2 statistic is inaccurate due to the fact that we ignored any spatial autocorrelation in the data, so we need to be cautious interpreting this result.) The urban/rural categories are the very largest cities (e.g., New York, Los Angeles, San Francisco, Chicago, and Philadelphia in the lowest row; Houston, Washington DC, Miami, Dallas, and New Orleans in the middle row) and all other HSAs. A band of HSAs in the Southeast has the highest percent of female-headed households, but these areas have

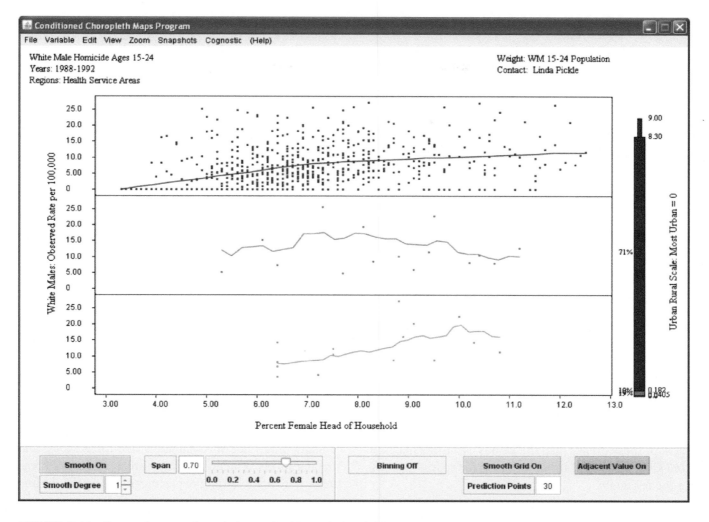

FIGURE 5.17 Scatterplot smoothes of observed white male homicide rates, 1988–1992, conditioned on percent female head of household.

homicide rates in all three rate categories. Except for separating out cities with high rates, no other patterns are obvious. Let's see if another type of view would clarify the associations among these three variables.

5.7.2 CONDITIONED SCATTERPLOT VIEWS

Scatterplots can provide a quick method to explore associations between candidate conditioning variables and the study variable. CCmaps optionally displays conditioned smoothed scatterplots instead of micromaps. Using the same slider settings as in Figure 5.16, the observed homicide rates are partitioned according to the row slider settings, here the urban/rural status of the health service areas (Figure 5.17). Among the most urban (bottom panel) and the less urban (top panel) places, there is an increasing rate of homicide with increasing proportion of female head of household; there is no apparent trend in the middle panel.

Several parameters can be set to adjust the degree of smoothing of the fitted line through the observed points (see controls at bottom of screen in Figure 5.17). Note that the range of rates displayed in each panel (0 to 27) is less than the range of the data shown in the top slider of Figure 5.16. This is because "Adjacent Value On" is selected in the lower right, meaning that the range is limited to the upper adjacent value. The top slider bar in Figure 5.16 lists this as 27.3 (the value defining the thin outlier segment of the bar).

Sophisticated modelers may also find CCmaps useful for model criticism by allowing the study variable to be the model residuals and the conditioning variables to be the residual standard errors or other variables that may turn out to be associated with the geospatial patterns. Looking for patterns in a model's residuals is a common analytic step to see if the predictor variables have adequately accounted for the spatial variation in the original data. Mapping model residuals can trigger

TABLE 5.2

Rank Ordering of State Homicide Rates, Categorized by Percent Female Head of Household, in Preparation for Constructing a QQ Plot

Values Based on Sorting States in Each Class		Class with Low Percent Female Head of Household		Class with High Percent Female Head of Household	
Rank	Cumulative Probability[a]	State	Homicide Rate	State	Homicide Rate
1	0.05	NH	2.0	DE	7.5
2	0.15	ND	2.1	NY	10.8
3	0.25	IA	2.7	GA	11.6
4	0.35	MN	3.8	TN	11.9
5	0.45	ID	3.9	SC	12.6
6	0.55	NE	4.1	NM	13.1
7	0.65	SD	4.3	AL	14.7
8	0.75	MT	4.7	MS	16.5
9	0.85	WY	5.2	MD	17.7
10	0.95	KS	6.4	LA	21.6

[a] (Rank – 0.5)/n.

ideas about the reasons for some of the extreme residuals, which can lead to additional model variables for better prediction. Those familiar with modeling may have good graphics they prefer for many tasks, but most of these do not map the values. Analysts may still find CCmaps useful to rapidly look for spatial patterns in the residuals.

5.7.3 CONDITIONED QQ PLOT VIEWS

The condition-based partitioning of regions into subsets provides the opportunity to compare study variable values across subsets. Many kinds of statistical comparisons are possible, such as the two-way map view in CCmaps that provides the mean of each subset. As indicated earlier, sometimes the subset means can be similar, while some variances differ. Different statistics may have different stories to tell. A visual comparison of the whole distribution for each of two subsets can sometimes bring out patterns that are related to the mean, the standard deviations, or additional distributional features. CCmaps provides an option to display conditioned quantile-quantile (QQ) plots (Wilk and Gnanadesikan 1968) for the visual comparison of distributions.

QQ plots are most often used to compare a data distribution with the normal distribution. In this case, curvature in the dot-connecting lines implies skewness and tail thinness. Some nonlinear patterns suggest that the distribution is a mixture of two distributions provided

that the pattern is supported by sufficient data. Several characteristics of the data can impact the shape of the QQ plot: the number of states included in each distribution by the slider settings, varying standard deviation of state values due to population size differences, and varying levels of aggregation of urban and rural areas to the state level. Strong departures from a straight line on the QQ plot warn us that our comparisons in terms of means and standard deviations are muddied and are not telling the whole story of the data.

Because many readers may not be familiar with QQ plots, we illustrate the construction in Table 5.2 with the state homicide data used in Figure 5.9. We formed two classes by choosing the states with the ten highest and ten lowest values of percent female head of household in 2000. We then sorted the states in each class by their homicide rates to obtain the state ranks within each class. New Hampshire and Delaware have rank 1 in their respective classes, so the algorithm pairs them for the QQ plot. North Dakota and New York both have rank 2 in common, so the algorithm pairs and plots their values. The QQ plot shows all the rank pairs with pairs connected by line segments in rank order (so the rank 1 pair is connected to the rank 2 pair, etc.).

QQ plot construction is a little more complicated when there are a different number of states in the two classes being compared. Instead of directly pairing states based on their rank, a quantile plot is constructed for each class where the pairs of points are cumulative probabilities and quantiles (the state values). The

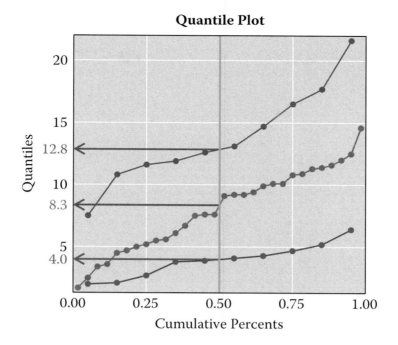

FIGURE 5.18 Cumulative probabilities for state homicide rates, males, 2001–2005, classified into low (ten states), middle (thirty states), and high (ten states) values of percent female head of household in 2000.

method for calculating the cumulative probabilities used here subtracts 0.5 from the rank and divides this result by the number of states in the class. The second column in Table 5.2 shows an example. Figure 5.18 superimposes the quantile plots for the low-, middle- (data not shown in Table 5.2), and high-value classes (blue, gray, and red, respectively).

With the superimposed quantile plot in hand we can take the next step that picks cumulative probabilities to use in a linear approximation in order to obtain cumulative probability-matched quantiles. Figure 5.18 shows how choosing the 0.5 cumulative probability results in the quantiles of 4.0, 8.3, and 12.8 for the three classes. We do this for a sequence of values typically bounded by the cumulative probabilities for the smallest class.

The quantiles based on states in any one CCmaps panel could be compared to the quantiles based on the states highlighted in each of the other subsets, but this would result in too many comparisons to display. Instead, CCmaps compares the distribution of a single panel to the distribution that includes the values of all other panels.

To illustrate the implementation of QQ plots in CCmaps, we return to the state homicide data. In Figure 5.19 we can see that the distribution of homicide rates for states with high percent female head of household (as defined by the sliders in Figure 5.9) is higher than for states in the low-value class. If the distributions

were equal, the points would fall around the equality line, i.e., where x = y. Since we are not a good at comparing slopes of lines unless they are close to a 45° angle, we have added a line indicating distribution equality. Most QQ plots zoom in to the range of the data and so are roughly square, helping to avoid extreme slopes in the plot. Clearly the points are far above the line. The plot shows that the mean for the high class is larger, not just because a very few values make the mean larger, but because all of the state values are high. If the standard deviations for the two classes were the same, the points would fall along a line roughly parallel to the line x = y. If we were to fit an approximate line to the blue dots, it would have a much steeper slope than the equality line. When the points are close to being on a straight line, the slope is approximately the ratio of the standard deviation for the two groups. Here the slope is far from 1, so there is a distribution mismatch in terms of both means and standard deviations. At least two points are departures from what appears to be a pretty straight line. Figure 5.19 suggests a mixture of distributions.

Figure 5.20 demonstrates the implementation of QQ plots as an alternative view to the maps in CCmaps, using health service areas in order to have an adequate number of units to display the conditioned QQ plot panels. There is a clear shift downward in the distribution of HSAs in the top left panel—all values are below the corresponding means in the other panels. The

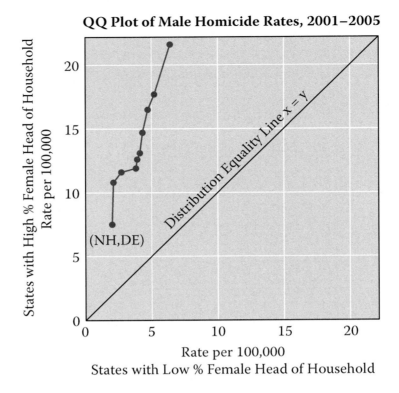

FIGURE 5.19 Comparison of the distribution of male homicide rates, 2001–2005, for states with high and low percent female head of household (as defined in Figure 5.9).

average of regions in this panel, 7.63 (shown in the box at the upper right of the panel), is the lowest of all the panels. The distributions of the other two top panels look similar to those of the other panels. The panels in the second and third rows contain only a few points, so nothing can be gained by their examination. For much larger data sets, such as for the three thousand U.S. counties, CCmaps limits the number of computed and displayed quantile pairs to one hundred per panel, using the cumulative probabilities for the smaller of the two sets up to one hundred values.

The conditioning sliders work in this alternative view of the data in the same manner they work in the map view. Clicking on a panel enlarges it in the same way clicking on a map panel does. Clicking in the enlarged view returns to the 3 × 3 view. The color boxes in the top left of each panel display the panel mean relative to the classes defined in the top slider and are updated in real time as the sliders are changed. CCmaps supports the use of region weights for distribution comparisons.

So far the map view in CCmaps is popular but the QQ plot view draws little attention. However, we hope that a better understanding of the power of this graphical display will lead to greater use.

5.8 CCMAPS SOFTWARE OPTIONS

5.8.1 Variable Selection

Our examples so far have used preselected conditioning variables. CCmaps includes a "variable picker" that allows you to quickly change the conditioning and study variables, as shown in Figure 5.21. The variable names are taken from the first line of the data input file. The variables to be used as a weight and to be displayed in the mouseover are also selected here.

5.8.2 Snapshots

Data exploration is not a linear process. Using CCmaps, we typically will try different views of the data, condition on different variables, and experiment with slider bar settings. Along the way, we may see particularly interesting views that we wish to return to later.

Screenshots can help to record the steps of our exploration, but it is not always easy to recreate these images. CCmaps facilitates record keeping by allowing the analyst to save live annotatable snapshots. These are not screen snapshots but are records of the state of CCmaps—variable choices, slider settings, and all. By specifying a live snapshot, the system returns to the

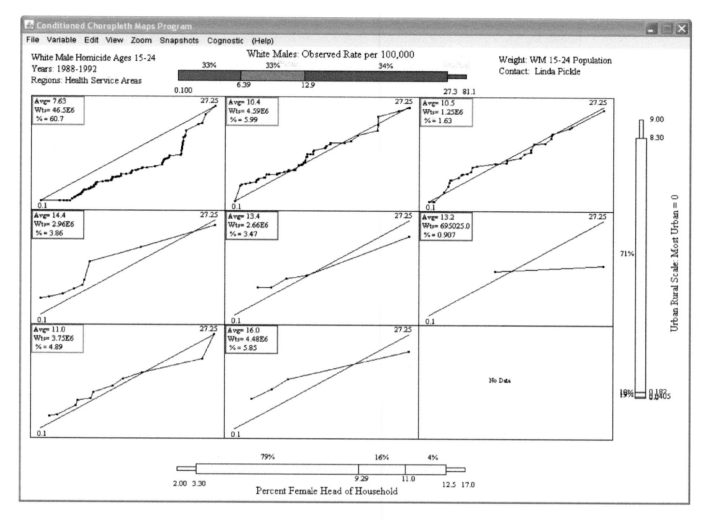

FIGURE 5.20 QQ plot view in CCmaps of male homicide rates by health service area.

settings when the snapshot was made, recreating that image exactly. The snapshot text annotation is hopefully enough to jog the analyst's memory about why this view was of interest. An experimental version of CCmaps includes voice annotation. An idea for future enhancement is to link to voice-to-text software so that notes could be saved without the interruption of having to type them.

In addition to helping to organize and replay an exploratory analysis, live snapshots are useful for preparing reports or presentations. Snapshots can be saved to build a sequence of results for later presentation. It is easy to jump from one view to another, such as from a map view to a scatterplot smooth, enlarge it, snapshot with comments, jump back to a previous snapshot, pick a new variable for a slider, find a good slider setting, make a live snapshot with comment, and so on. The annotations can be edited and the snapshot order reorganized for live dynamic presentations.

5.8.3 DISPLAY OPTIONS

In addition to the usual options of pan and zoom, color choices, and mouseovers, CCmaps allows control over boundary line thickness and toggles internal boundaries on and off. Each variable can be scaled by a different factor, and the corresponding slider labels can be altered to display the new units.

Clicking on one of the micromaps will enlarge that one image to the full screen. Zooming in to a subregion of the micromap in this view will change all nine micromaps to the same map extent. Clicking on the map image again returns to the 3 × 3 view.

5.9 SUMMARY

Conditioned micromaps facilitate thinking about three variables at a time in a spatial context. By conditioning on two variables at a time, this is a more sophisticated tool than linked micromaps. Although linked

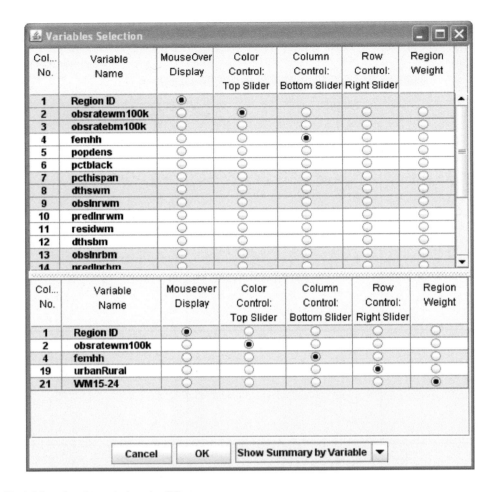

FIGURE 5.21   Variable selection window in CCmaps.

micromaps can present statistics for multiple variables, the variable values are displayed in separate columns, whereas the conditioned micromap displays a single variable on the maps conditioned on two other variables. The conditioned micromap is more of an interactive, analytic tool. For example, an epidemiologist may speculate about the association between two columns of statistics presented in a linked micromap format, but can explore this hypothesized association by examining spatial patterns of one variable while controlling for the other using the conditioned micromap format.

The CCmaps software used in this chapter is available for free download on our website, along with region boundary files and data sets for all of the examples. Future enhancements include adding the ability to import boundary files and construct study-specific boundaries, the ability to export display images, and

other statistical tools for pattern recognition, such as SaTScan (Kulldorff et al. 2006) to identify significant clusters on the map.

The ease of use of CCmaps can provide an educational pathway for some users, combining maps with statistical graphics and statistical feedback, drawing them toward more sophisticated analyses or models. The data exploration by both novice and experienced analysts can be facilitated by CCmaps' data analysis guidance and management features, such as the cognostics and the live snapshots. We have used CCmaps to illustrate the integration of commonly used statistical procedures with georeferenced graphics. However, we encourage other software developers with access to greater statistical and mapping functionality to focus attention on this area. Our goal is to facilitate data exploration and to promote the development of educational pathways in software.

6 Comparative Micromaps

In learning to read comics we all learned to perceive time **spatially**, for in the world of comics, **time and space** are **one and the same**. The problem is **there's no conversion chart!** A few centimeters which transport us from **second to second** in **one** sequence could take us a **hundred million years** in another.

—Scott McCloud (1993)

6.1 INTRODUCTION

In this chapter, our final stop on the tour of micromaps, we propose designs for a sequence of maps, typically over time, that we call comparative micromaps. This type of micromap promotes comparisons among one- and two-way indexed sequences of complete maps.

McCloud's book, *Understanding Comics* (McCloud 1993), provides considerable practical guidance that is, perhaps surprisingly, relevant to sequences of quantitative graphics. Our graphics are caricatures often made possible by the power of statistics to provide terse descriptive summaries. Some of our statistics are about specific facets of billions of person-years of human experience. These statistics are tiny caricatures of the richness and drama of human life. Our maps are caricatures too. These very simplified maps can be informative as long as they convey region identity and relationships to neighbors. When a series of micromaps displaying statistical summaries are placed side by side to illustrate changes, we often have difficulties seeing everything that changes from one panel to the next. As we explained in Chapter 2, this is usually due to change blindness, i.e., our mental image of one map is wiped out before we can capture the next one, and to a limited visual working memory that can only hold a little information about no more than three very simple objects. Animation of complex images is even more difficult because our mental comparison must be based on the last image that is no longer visible. A recent study demonstrated the superiority of small multiples over animation for fast and accurate map comparisons (Robertson et al. 2008).

By thinking about a series of micromaps as side-by-side graphical caricatures that step from left to right over periods of time, you can see why the design of comics might be relevant. Comics include text as "bubble comments" to help explain what is happening from one panel to the next. We adapt this concept for comparative micromaps to call attention to what is changing. Our bubble comments take the form of a shifted sequence of juxtaposed micromaps that displays the location and nature of the changes explicitly, removing the cognitive burden of mentally calculating differences from the reader. This distinguishes comparative micromaps from a standard time series of maps displayed side by side.

The comparative micromap design is the newest type of micromap plot. The design is based on principles outlined in Chapters 2 and 3, modified by our experience and results of usability studies with other micromaps. Direct evaluation of comparative micromaps may yield new insights. We hope that presentation of our initial ideas here will stimulate your thinking about micromap comparisons that might be useful for your own data.

We created the new examples in this chapter using customized R functions and scripts that are provided on the book website. The legends in our new examples are similar to the dynamic sliders in the Java implementation called TCmaps created by Carr and Zhang in 2006. This implementation is evolving to incorporate recent two-way and second difference designs. Usability tests and further reflection may lead to evolution of the layouts and dynamic sliders, but we anticipate changes being modest based on our previous experience. TCmaps screen shots in Section 6.6.1 indicate additional features not available in our R scripts such as horizontal scroll bars for long time series and alternative views.

Chapter 1 introduced a simple example of comparative micromaps that we repeat in Figure 6.1 with annotation. This is a sequence of four U.S. state unemployment maps indexed by time from 2001 to 2004. The design assigns the states to one of three classes—low, middle, and high—based on their annual values. The second row of maps in the figure highlights the states that have changed class membership from one year to the next. It shows the states in the color for the class to which they change. This is one of several ways to represent change. After we address map indexing and transformation for comparability, we will return to the topic of calling attention to change, discuss displaying first

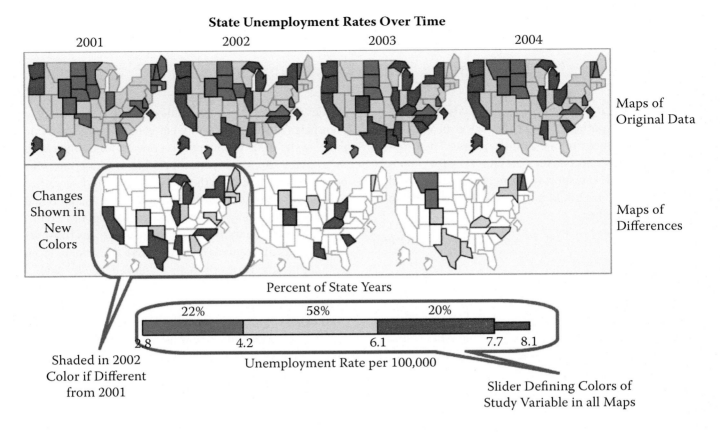

FIGURE 6.1 Comparative micromaps of annual state unemployment rates, 2001–2004, illustrating the basic design, annotated to show basic components.

and second differences, and then end the chapter with alternative views to the simple design of Figure 6.1.

6.2 REPRESENTING CHANGE—ONE-WAY COMPARISONS

6.2.1 LAYOUTS FOR UNORDERED CATEGORICAL VARIABLES

Time is not the only way to index a sequence of maps. Sometimes we use ordered categorical variables such as age groups to index a map sequence. We can also index maps with unordered categorical variables such as sex or race/ethnicity, although we then need to impose an order on the categories to create a sequence of maps for display purposes. The imposed order leads to paired comparison issues when there are more than two maps. With three or more maps in a linear sequence, the pairs of maps we want to compare may not be adjacent. With a sequence of maps for just three categories, such as non-Hispanic white, Hispanic, and non-Hispanic black, we can repeat the first map as a fourth map at the end of the original series. This permits side-by-side comparisons of every pair in

the sequence: non-Hispanic white versus Hispanic, Hispanic versus non-Hispanic black, and non-Hispanic black versus non-Hispanic white. This does not generalize to more than three unordered categories.

A common statistical approach for three or more unordered categories is to make comparisons of each group to a composite of the groups. For example, statistics for each race/ethnicity group can be compared to the statistic for the combined population. This leads to layout possibilities such as showing a combined micromap with a sequence of three group-specific difference micromaps below. With four categories, the difference micromaps could be placed on the four sides of the composite micromap. With six categories we could frame micromaps with hexagons and put six micromaps in a hexagon circle around a center hexagon-bounded micromap that is the common reference. In fact, we have recommended this circular design for a public website that asks people to select simulated galaxies that reasonably approximate a real galaxy shown in the center (mergers.galaxyzoo.org). The site enables humans to teach a computer how to set galaxy simulation parameters so that the simulations are very similar to a growing

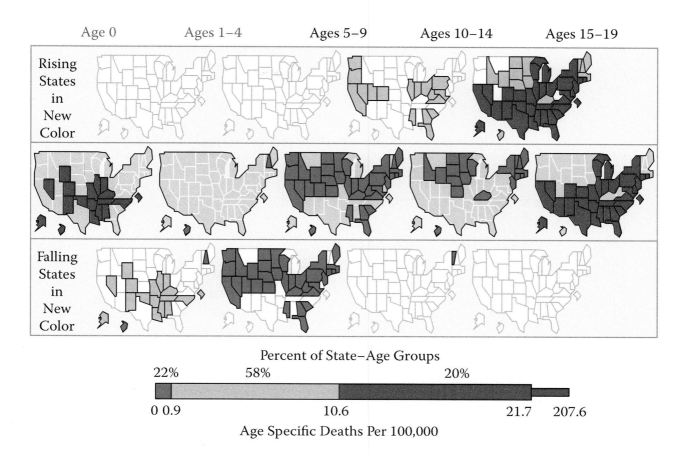

FIGURE 6.2 Design highlighting different interval widths for the micromap series—state homicide rates among males ages 0 to 19, 1996–2000, stratified by age group.

list of known galaxies. The computer will learn from the human guidance using a genetic algorithm. This idea was inspired by the Galaxy Zoo project that now has over two hundred thousand people classifying features of known galaxies in order to find consensus of the classifications or unusual features that have not yet been explained (Lintott et al. 2008). Many facets of comparison designs, such as page layouts, are not restricted to micromaps, and there are good layouts that could be borrowed from other fields of study.

6.2.2 LAYOUTS FOR ORDERED CATEGORIES

The designs in this chapter focus on one- and two-way linear sequences of ordered categories. For times series we assume regularly spaced time periods and so our basic design has the micromaps equally spaced, as in Figure 6.1. Alternative layouts for irregularly spaced time series might show the spacing based on gaps between the micromaps. In Figure 6.2, micromaps of homicide rates among young children show a somewhat different irregularity. The infants (age 0) have been separated from the first age group in a sequence of five-year

age groups. The micromaps show why—homicide rates are much higher for infants than for boys age 1 to 4. We have highlighted the nonstandard age intervals in red. There are many other ways to deal with irregularities that must be evaluated with space constraints in mind.

Most examples in this chapter show a horizontal sequence of micromaps with the time or other index increasing from left to right. Guidance concerning orientation and order is often based on following the dominant convention for the targeted audience or culture. Some well-established conventions conflict. For example, the table-reading convention is to read top down, while the graph convention is to read increasing values from the bottom up. The next example is in a vertical orientation with time increasing from the top down. Of course, the latitude coordinates in the maps shown increase from the bottom up. This is a case where those in the southern hemisphere may prefer using the table convention for maps. Yes, a change in convention can turn the world downside up.

Figure 6.3 displays yield spreads (ten-year versus two-year bond interest rates) for government bonds in

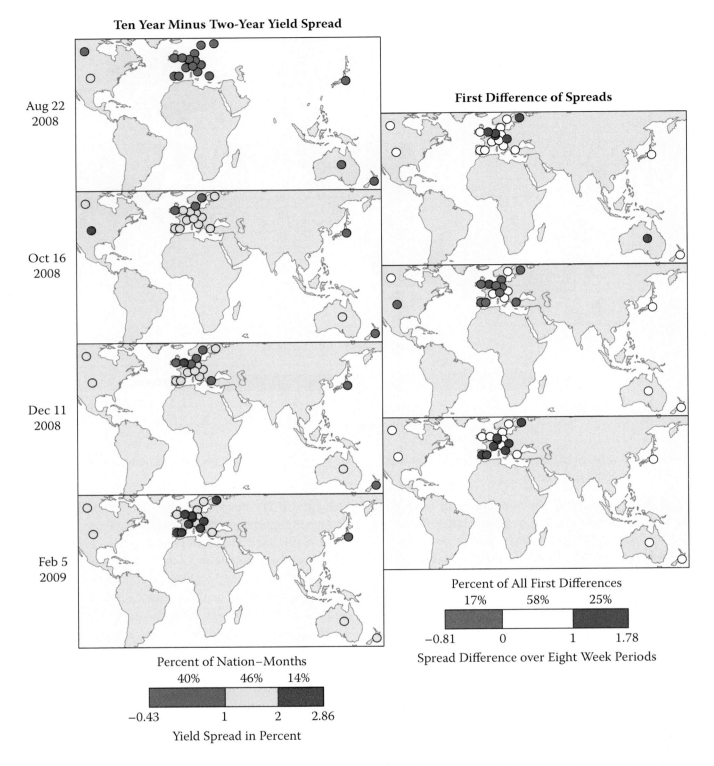

FIGURE 6.3 Comparative micromaps in vertical orientation, using color-filled dots instead of color-filled nation polygons to represent the ten-year minus two-year government bond yield spread, August 22, 2008, to February 5, 2009, by eight-week intervals.

nineteen countries over a period of world financial crisis, August 2008 to February 2009. The world map has been modified for a more space-efficient display. Much of our data were for European nations, so we cropped a map of continents and islands on all four sides to increase

the distance between the centroids of these nations. Further modifications included moving New Zealand and Iceland and removing part of the Atlantic Ocean. This modified map is still relatively wide, so in order to display four maps on a page, a vertical orientation is used. The difference maps are calculated by simple subtraction of yield spreads of one map from the previous map, requiring a separate slider for these values. These maps are offset between the yield spread maps, as in Figure 6.1, but there is no space between the maps to maximize the usable space. Because there are over one hundred countries in the world with very variable geographic areas, we used a constant-sized dot for each country, color coded for the yields or their differences, categorized by their respective sliders. One design challenge was choosing a symbol size large enough that its interior color would be visible but small enough to limit overplotting. In addition, bold, saturated colors are used to maximize visibility of the small symbol colors.

Yield spreads are considered to be a forecasting (leading) indicator of a nation's economic health. A negative spread, where the longer bond has a lower rate than the shorter one, is rare and usually predicts a recession. Economists watch changing yield spreads to try to predict near-term economic changes. Increasing yield spreads can predict inflation, if the increase is due to increases in the longer bond rate, or a "business-friendly" federal fiscal policy when the government reduces the shorter bond rate in an effort to stimulate the economy.

Newspaper articles over the fall of 2008 suggested that the United States had preceded the rest of the world into a major recession, so we thought we would explore some relevant data using comparative micromaps, looking for evidence consistent with or contrary to these claims. Figure 6.3 shows that the United States was the only country of the nineteen that had a spread as high as 1 to 2% on August 22, 2008; others had lower spreads, with negative spreads in New Zealand, Sweden, and the UK. Eight weeks later, the spread in the United States had widened to over 2%, while most other countries had risen to 1 to 2%. For the last two dates, the U.S. spread lowered back to 1 to 2%, while many other countries had continuing increases in their spreads. The first difference maps show this more clearly, especially between the last two dates, when seven countries had increases of more than 1% while the U.S. spread increased by less than 1%. The changing U.S. spreads during this time were due to declines in the two-year bond rate from August to October, declines in both rates from October to December (when the ten-year bond rate was cut in half to 2%), and increases in the ten-year bond rate from December to February, returning to their previous 4% level by June 2009. These micromaps suggest that the sharp increase in the U.S. yield spread in October presaged a worsening economy, but that the return to a more normal spread by February indicated that the worst was over, while many other nations continued to worsen. Although categorization of the spreads has reduced the information in these data, the patterns are consistent with our deepening recession over the 2008–2009 winter and an apparent bottoming of the U.S. stock market in March. Time will tell whether the rest of the world will soon start to recover.

6.2.3 Representing Change

As we examine the first row of micromaps in Figure 6.1, we need to look back and forth, mentally doing pairwise comparisons to judge what has changed. This is difficult even at the state level because of our visual and cognitive limitations. As discussed in Section 2.4.3, we are essentially blind when our eyes jump to a new image because the old retinal image has faded away in about 1/5 of a second. Furthermore, only about three simple objects can be retained in visual working memory long enough for us to make comparisons. This is why we have to look back and forth so often to compare images and then are mostly comparing different small focal areas within the larger images.

Animation has been a common method of displaying changing map images. Astronomers successfully used aligned alternating photo images to identify an object's changed position in a field of stars and galaxy that remained in the same fixed position. The moving object would trigger our visual change detectors and appear to blink. This can work well if only a few areas blink to indicate a change, or if there is a single location that the viewer can focus on, such as the spread of an epidemic from a single focal point. Google Earth™ has popularized animation of maps, showing changes over time and as the viewer "flies" over the region. This is a fun tool but is not intended to be an analytic tool for statistical data. An early study of the effectiveness of animating disease maps found that some people used a slow-speed animation but others turned off the animation and carefully stepped back and forth through the time sequence (MacEachren et al. 1998). Still, some of the geographic changes were missed by the viewers. Why not just show the change in maps directly so people can point to changes and talk about them?

As you can see in Figure 6.1, more than a few states change unemployment categories each year, so animation is unlikely to be effective for these data. These problems are much worse for sub-state-level micromaps (e.g., counties or census tracts) because there are so many

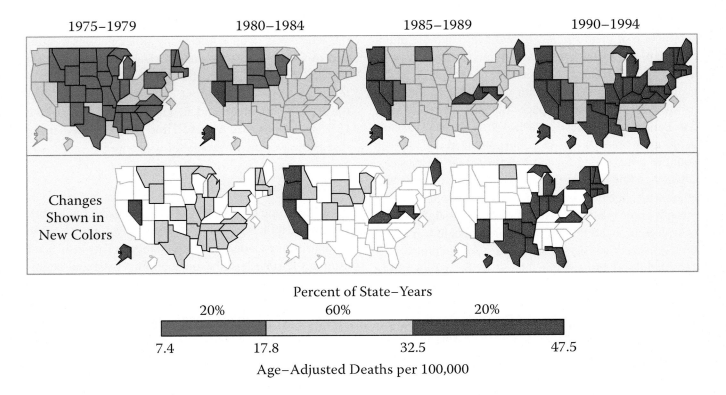

FIGURE 6.4 White female lung cancer mortality rates by state (age adjusted to 1970 standard), with change indicated by the new color for each state.

more color-coded regions to observe. Showing a map for each time period (or other categorizing variable), juxtaposed closely in a single row, allows the best opportunity for communicating change accurately.

Many of the data sets of interest to us have data about hundreds to thousands of regions. We find county-level data more interesting than state-level data because the latter hide so much local variation. However, there is value in studying maps at different scales. Small area maps are so complex that we still would have trouble evaluating the changes from one map to the next due to the cognitive limitations mentioned above. This leads to the second row of the comparative micromap design, where the micromaps now indicate the change explicitly. These micromaps are offset from the first-row maps and so convey to the reader that these are difference maps, i.e., between the upper two maps. Figure 6.1 on unemployment rates simply color codes any state that changed class by its new color class. We see that most of the change, at least as measured by a change in category, occurred from 2001 to 2002—seven states moved into the highest category and ten states moved into the middle category as the United States began a recession. Fewer states changed from 2002 to 2003 and unemployment started to decline by 2004 as the U.S. economy improved.

Most of the time, change occurs when a region moves from one category to the adjacent one, but the less frequent jump from low to high or high to low can be interesting. Two limitations of the simplest design shown in Figure 6.1 are that this method of color coding on the change micromaps tells us what the states changed to, but not what they changed from, and not the magnitude of the change. We can look at the preceding map to resolve the question, e.g., all of the states that moved into the middle category in 2002 had been low in 2001. However, this involves a level of scrutiny that can become tedious. An alternative color scheme for the change micromaps explicitly codes the direction of change.

We illustrate this alternative display by examining the dramatic changes that occurred in white female lung cancer mortality rates over a twenty-year period (Figure 6.4). Most state rates were below 17.8 per 100,000 in 1975–1979 but over 32.5 per 100,000 in 1990–1994, nearly a doubling in less than twenty years. Further, we can see that these changes were not uniform geographically. Rates along the West Coast rose earlier—this new cluster of higher lung cancer rates among white women was first noticed on a 1970s' atlas map (Pickle et al. 1987) and expanded eastward during the 1980s (Pickle et al. 1996).

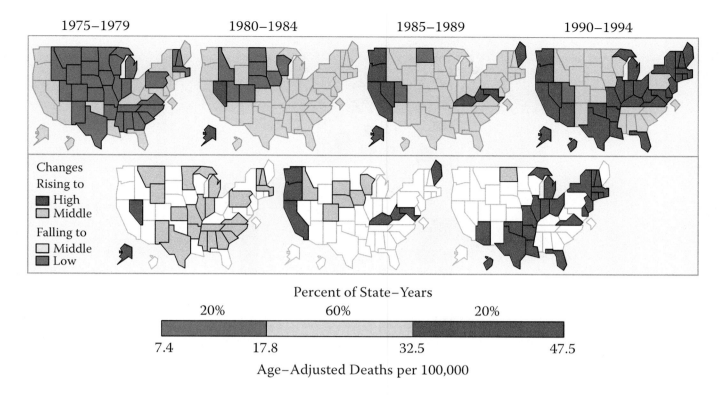

FIGURE 6.5 White female lung cancer mortality rates by state (age adjusted to 1970 standard), with change indicated by a bivariate color scheme.

By looking only at the change micromap row, we might wonder if the states shaded gray on the first map had risen from the low category or dropped from the high category. The modified color scheme in Figure 6.5 makes this clear. Now we see that every class change over twenty years was a rise in rates—red indicates a change from low or middle to high, and pink indicates a change from low to middle. Similarly, light and dark blue would indicate falling rates, had there been any.

This is a compact design, but some readers might not immediately understand the bivariate color scheme on the change micromaps. Alternatively, rising and falling rate changes can be shown on separate change rows. Figure 6.6 presents all cancer mortality rates for white women for a forty-five-year period. In this three-row design, the upper change row indicates that rates were rising between the periods and the bottom change row indicates that they were falling. The color in the change rows is again the color of the category to which the state changed. We easily see that the all cancer mortality did not change much until the late 1970s, when they started to rise in many states (all change is in the upper row). Rates began to fall in the last two time periods (all change is on the bottom row).

Figure 6.6 demonstrates how much information can be displayed by the comparative micromap design—nine

time periods are shown. This may seem to be too compressed for easy use, or you may have an even longer time series to display. What are other display options? First, because comparative micromaps are primarily for exploration, we are more likely to use a computer display and so are not limited to the size of a single printed page. Hardware solutions provide one set of options. With two large LCD screens for a working environment, it is possible to have an available width of 40 inches or more for a display. Software and design solutions provide another set of options. For example, TCmaps uses a scroll bar to show a long series in chunks on the screen. If your space is really constrained, say for a print publication, the long micromap series could be split into two long panels that are presented one above the other. Using Figure 6.6 as an example, the first five rate micromaps and their corresponding difference maps could be displayed across the top of a landscape-oriented page. To ensure continuity, the rightmost micromap (fifth of nine) on this top row would be repeated as the leftmost micromap of the bottom row, to give context to the first difference maps on the bottom row. If none of the above options are palatable, you can always display a subset of the longer series, e.g., every other time period.

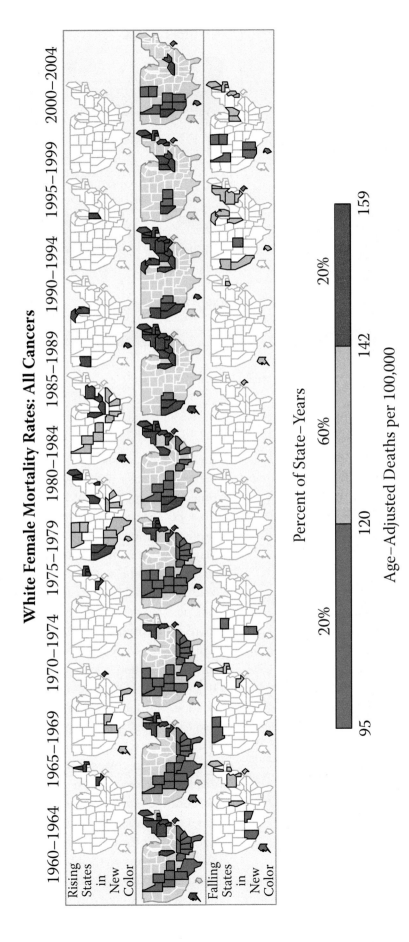

FIGURE 6.6   Rates of mortality among white females due to any malignancy, 1960–2004, by state, with changes shown separately for rising and falling rates.

6.3   TYPES OF COMPARISONS

Comparisons provide the foundation for description and discovery. There is an old joke about a parent asking his or her college sophomore what he or she had learned from the first year at school. The reply was "Compared to what?" A key task in the production of comparative micromaps is to provide a coherent basis for making comparisons. We must choose the framework for making comparisons unless the original data are inherently and easily comparable. Different comparison frameworks bring out different patterns. Our frameworks may be based on a reference distribution, single reference value, or sequence of reference values. They may be local or global. The criteria for creating classes may be based on the percent of regions in the class, the percent of region weights (such as population size) in the class, thresholds currently thought to be important for interpretation in the application area, or hypothesis tests indicating significant differences that will ignore some of the changes that may well be due to noise.

The examples so far have used the global state-year value distribution in the slider, i.e., the distribution of all the state values plus Washington DC values over the entire time span shown, to define three class categories for all of the maps. (We have adapted this term from the epidemiologist's person-years concept, where a person is counted in each study year that they were alive. This avoids biasing the calculation by counting as equal persons who lived many years versus those who only lived a few.) In Figure 6.5, which has four time periods, the distribution has 51*4 = 204 state-year values. We sought to put roughly 20% of the state-year values into the low and high classes. If there is a time trend in the data, we should see the colors shift over the time periods. Figure 6.5 provides such an extreme example, with mostly blue shading in the first period to mostly red shading in the final period.

Just as for the conditioned choropleth micromaps in Chapter 5, the sliders facilitate this by dynamically updating the percent of observations that appear above the midpoint of the three color bars in the slider. The micromaps lift the low and high classes into the foreground using color and so focus attention on a more modest number of states. If there is some spatial coherence, and often there is, the patterns can encourage thought. We can be intimidated and move on to something else if it appears that there are too many regions or too much variation to deal with at one time, so focusing on subsets of the data defined by the slider can make the task of seeking patterns more manageable.

It is important to keep in mind that there is only a single slider bar defining the color categories for all of the maps. If the interpretation of the color red changed from map to map, e.g., if red indicated the highest 10% of rates on that map regardless of the actual rate values, the interpretation of the sequence would be more complicated. The use of a constant rate categorization method over all maps to be compared in a series has been found to work well (Brewer and Pickle 2002). The NCI atlas website allows the user to request rate categorization using either the relative (each map separately) or constant (based on total rate distribution) method for nine 5-year time periods (1950–1994).

Because the values that we have used in these examples, i.e., directly age-adjusted rates, have no meaning except in comparison to other similarly adjusted rates, mortality atlases tend to categorize rates by some quantiles of the distribution for each map separately, e.g., decile categories in the last NCI atlas (Devesa et al. 1999) and a combination of deciles and quintiles in the NCHS atlas (Pickle et al. 1996). Therefore, most epidemiologists are used to interpreting the colors on each map in a relative rather than an absolute way. However, interpreting micromaps with percent-based slider thresholds, the specified percent of regions is forced into the highest and lowest classes, even if there is a narrow range of the original values. For some applications, we may prefer to set the thresholds by the actual values rather than the percents.

We can incorporate time-period-specific reference values in our construction. For example, we could subtract the U.S. average mortality rate of each time period from the values of all regions on that map and use those differences in the global distribution represented by the slider. Alternatively, we could divide the state rates by the U.S. rate to obtain the comparative mortality ratio (see sidebar for more details). These transformed values could be used as the global distribution for the slider categorization. This latter method was used in the NCHS atlas to put maps of every cause of death on the same scale; then all maps were presented on a single page for direct pattern comparison (Pickle et al. 1996). A five-class color scheme was used for display of these relative difference maps, with rates ±15% of the U.S. rate considered to be roughly equal to the U.S. rate, and rates more than 25% discrepant from the U.S. rate considered to be outliers.

Of course, we are not restricted to making map comparisons based on differences. Comparisons based on ratios are sometimes preferable, particularly when

Age Adjustment

Throughout this book we have used the term *age-adjusted rate* to implicitly mean rates adjusted for age differences in populations using the direct method of adjustment. The directly adjusted rate is a weighted average of age-specific rates in the population of interest, weighted by the proportional age distribution of the standard population (dos Santos Silva 1999). We occasionally have used the 1970 U.S. population as our standard population in order to be consistent with data from the 1999 NCI cancer atlas (Devesa et al. 1999), but unless otherwise specified, we have used the 2000 U.S. population as the standard. Adjusted rates are not meaningful per se because their values depend on the standard chosen, but directly age-adjusted rates are comparable if they were adjusted using the same standard.

The *comparative mortality ratio*, also called the comparative mortality figure, is a relative measure of mortality computed by dividing each directly adjusted rate by another comparably adjusted rate for the same time period (dos Santos Silva 1999). For our examples, we compare each region's rate to the U.S. rate. This ratio puts otherwise disparate rates on the same scale, i.e., the percent difference of each region's rate from the U.S. rate.

Many prefer to use the indirect method of adjustment, which results in the standardized mortality (or morbidity) ratio (SMR). Reasons for this preference include (1) it is easily interpretable as the number of excess cases expected in a region compared to the expected number in the standard larger region; (2) because rates from the standard region are applied to each smaller region, this measure will be less variable than the direct method when region populations are small; and (3) it can still be calculated if only the number of cases in each region is known but the population is not. However, the SMRs are only computed to be comparable to the standard population and are not directly comparable to each other unless certain data requirements are met (Pickle and White 1995). For example, a region with an SMR of 4 does not necessarily have twice the risk of a region with an SMR of 2; this region-to-region comparison depends on the corresponding population age distributions. Therefore, vital statistics researchers have recommended the use of directly standardized mortality rates. Because our comparative micromaps design involves subtraction of rates from two time periods, with potentially two different sets of age distributions, we do not recommend the use of SMRs for this visualization. This is an area requiring further study.

comparing values that are not on the same scale. For example, rather than comparing mapped rates for males and females, which often have quite different distributions, we could first calculate the female:male ratio for each region and apply the design of Figure 6.4 or Figure 6.5 to this ratio measure. We could go one step further and calculate the ratio of female:male comparative mortality ratios, really a ratio of ratios, and display these in comparative micromaps, as we have done in Figure 6.7. Note that there are gaps between the time periods displayed; the regular skipping of plots allows the display of a longer time span with a modest number of micromaps.

We know (see Figure 6.8) that female lung cancer rates on the West Coast have been higher than the U.S. rates (as measured by the white female comparative mortality ratio). We can see from Figure 6.7 that the white female ratio has been consistently greater than the comparable white male ratio; the opposite is true in southeastern states. That is, the relative measure compared to the United States is higher for women even though their actual rates are lower than for men. This may be due to differences in tobacco use by women in different U.S. regions—southeastern women were known to prefer to use smokeless tobacco in the past, which led that area to have lower lung cancer but higher oral cancer rates

(Winn et al. 1981). Regardless of whether differences or ratios, with or without reference value adjustment, are used, we expect that small population states will vary more around the class thresholds because their rate estimates have large standard errors. Some stabilization of patterns and trends can be achieved by aggregating to longer time periods, e.g., mapping average annual rates over ten-year instead of five-year periods.

6.4   TWO-WAY COMPARISONS

Epidemiologists who regularly calculate and compare relative risks and risk ratios might be comfortable thinking about comparative micromaps in terms of differences between rows of ratios, as in Figure 6.7. Other analysts may prefer to retain the original scale of the data because of its familiarity. How can we extend the basic comparative micromap design to permit comparisons across a second categorical factor, such as gender?

Class change panels can be interposed between adjacent cells of a two-way grid showing comparable values over time and across the two categories. In the example shown in Figure 6.8, white male and female lung cancer mortality rates were first converted to comparative mortality ratios to put them on the same scale; then these were mapped for three 5-year periods. Because the

Lung and Bronchus Cancer
Ratio of White Female to White Male Comparative Mortality Ratios

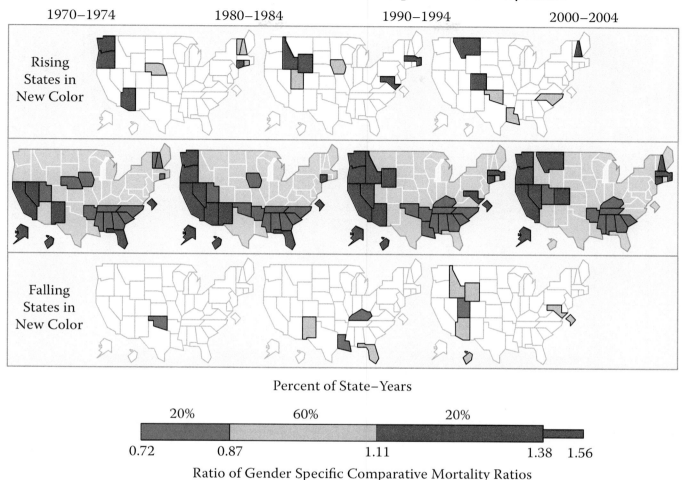

FIGURE 6.7   Comparative micromap design for ratios of white female lung cancer comparative mortality ratio to corresponding white male values.

design must now include a row of micromaps for each category, here gender, the change micromaps have been placed between the time-specific micromaps rather than offset on a second row as before. The different background shading identifies the type of micromap in that panel—the light blue-green panels are the original data and the gray panels are the change micromaps. Likewise, comparisons between the male and female micromap rows are shown as difference panels between the corresponding time-gender-specific micromap panels. In this example we use the four-color design introduced in Figure 6.5. The class change panels show states rising to higher classes (going left to right or top down) in pink and red. The pink states are rising to the middle class, and red states are rising to the high class. The class change panels show states falling to lower classes in light blue and blue. The light blue states are falling

to the middle class and blue states are falling to the low class. The low class (blue) has the lowest comparative mortality ratios where the comparison is to the U.S. rate for the five-year period.

The class change panels for comparisons over time are relatively empty, but the class change panels between females and males highlight many mismatched classes. With these class thresholds and global scaling, it appears that regional differences for women have dissipated over time. The male clustering of high rates in the Southeast has expanded by three states that are easy to spot as red states in the two class change panels (Figure 6.8, bottom row, maps 2 and 4) and the shifting of the blue cluster is made precise by noting the pink and blue regions in the two change panels.

As indicated earlier, the class change encoding with four colors leaves something to be desired. The

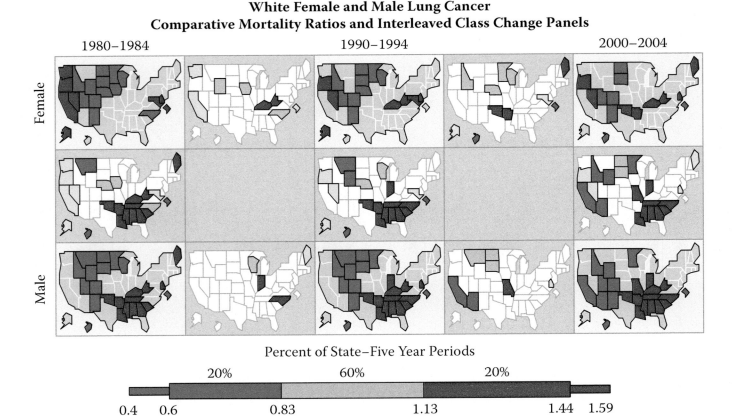

White Female and Male Lung Cancer
Comparative Mortality Ratios and Interleaved Class Change Panels

FIGURE 6.8 White female and white male lung cancer mortality rates, compared over both time and gender. Interleaved class change panels show the new class color (read left to right and top down) for the blue and red classes. Changes to the gray class are recoded as pink if increasing and light blue if decreasing.

rise from low to high and middle to high has the same color encoding, i.e., red. Similarly, the fall from high to low and middle to low has the same color encoding, i.e., blue. There are several options to rectify this situation, such as introducing two more colors (that fortunately will rarely be used). If the maps were a little larger we could plot a dot on the state to flag an extreme change. The color encoding is also subject to misinterpretation. Some readers may think of light red as being the average of the colors for the two populations or time periods being compared. Unfortunately, this average encoding loses the directionality of the change and is not recommended for this visualization application.

Many epidemiologists will want to delve into comparisons more deeply than just noting that a state has changed classes from one time period to the next. For states with values that are near a class threshold, the change could be very slight from one side of the threshold to the other. Conversely, the values could change substantially within a class and not appear as having

changed. The simple appearance of a static three-class map is bought at the price of loss of detail, sometimes important detail. The use of a dynamic slider can help some, allowing us to check the sensitivity of the micromap patterns to slight changes in the threshold levels. Cognostics, as discussed in Chapter 5, can greatly expedite the search for the patterns. Alternatively, we can provide more detail about the differences by using a five-class color scale. In Figure 6.9 we show difference micromaps instead of the class change micromaps in Figure 6.8. These maps show the differences by states, i.e., by subtraction, either between time periods or between males and females, partitioned into five 20% classes that are encoded by color. Epidemiologists who have studied the disease in question may have a good grasp of what differences or rates are of practical importance in terms of taking action. They may want to set the class boundaries to their practical decision-making thresholds rather than let the percents and data determine the boundaries. Then they are in a better position to act

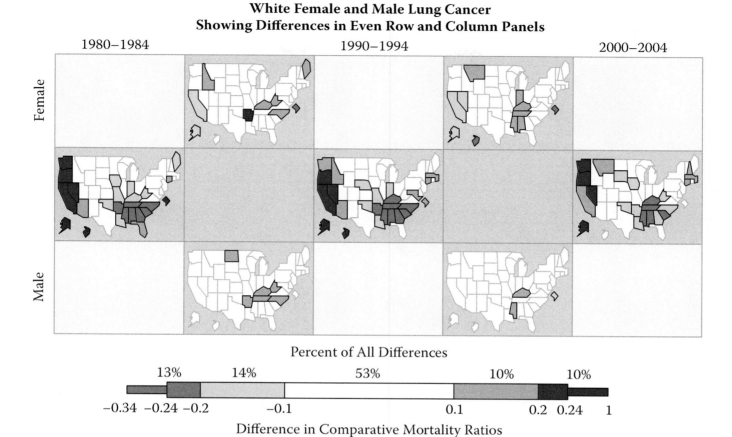

**White Female and Male Lung Cancer
Showing Differences in Even Row and Column Panels**

FIGURE 6.9   Interleaved class change panels from Figure 6.8 replaced by panels showing comparative mortality ratio differences encoded using five color classes. Differences are right panel minus left panel and bottom panel minus top panel. This particular design leaves states with small differences in the background.

based on the patterns that appear. The greater complexity of this five-class scale is offset by the greater detail it provides. Comparative micromaps provide alternatives that can be selected based on the task and audience.

6.5   RATES OF CHANGE

We can think about the states from many perspectives, such as the mortality rates, the change of mortality rates, i.e., the trend from one time period to the next, and the change in the trend over time. In physics terms, we can think of these three perspectives as location (the static values), speed, and acceleration/deceleration. We use this physics analogy to think about the forces associated with the acceleration and deceleration of mortality rates.

A concrete example might make this clearer. Suppose your son is driving from college to Florida for spring break. Because you are concerned about his safety, you monitor his location every few hours by a GPS installed in his car. At each time point, the GPS

beams the location of the car back to your computer. By subtracting the locations to get the distance traveled and then dividing by the length of the time interval, you can calculate the average speed he is driving during each segment of the trip. You are happy to see that he appears to be driving within the speed limit for most of the time. By subtracting the speed of one segment from the previous segment, you can calculate whether he is speeding up (accelerating) or slowing down (decelerating). You are not surprised to see that he accelerates when he nears his destination, anticipating the good times.

The top row of Figure 6.10 presents the same lung cancer data that we saw in Figure 6.5. In the physics analogy, these are the static values or locations. Unlike the previous layouts, this row is supplemented by rows of micromaps showing first differences (second row) and second differences (third row). The second row is conceptually the same as the second row in Figure 6.5; i.e., it maps the results of subtracting the state rates from one time period to the next, which we call the trend

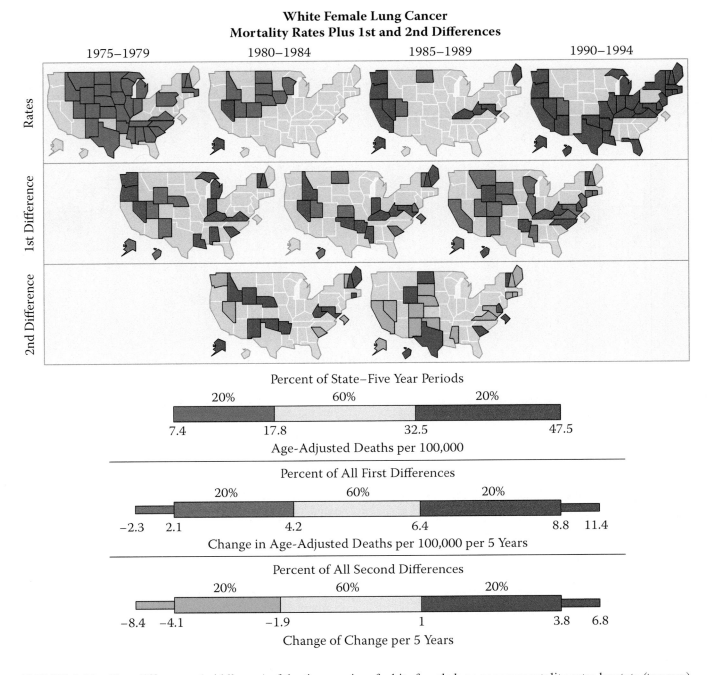

FIGURE 6.10 First differences (middle row) of the times series of white female lung cancer mortality rates by state (top row). Bottom row displays the second differences (differences of the second row).

(speed) of the rates. However, now these change maps have their own slider bar with associated cutpoints. Here we set the slider to color code changes that were in the highest or lowest 20% of all the state changes by brown or blue-green, respectively. The third row comprises maps of differences of the second row maps that display the degree of acceleration/deceleration of the rates, i.e., the changes in trends. These too have their own slider bar that categorizes the mapped values into the highest and lowest 20% of all of the state second

differences. The values on the sliders are important to the interpretation—negative values in the bottom slider are indicative of decelerating rates.

What does this all mean? Let's focus on the rates in California, which we show in Table 6.1. For simplicity, we will refer to the time periods as 1980, 1985, and 1990. The population of this state is large, so the patterns should be stable and reliable. From the top row of Figure 6.10, which contains the type of mortality maps that are usually published, we see that California rates

TABLE 6.1
White Female Lung Cancer Mortality Rates for California and Their First and Second Differences from Figure 6.10

1980–1984	1985–1989	1990–1994
28.4	33.4	35.8
5.0		2.4
	−2.6	

went from the middle category (rate = 28.4) in 1980 to the high category (rate = 33.4) in 1985, and then stayed in the high category (rate = 35.8) for the last period. The second row shows that the rise in rates from 1980 to 1985 (33.4 − 28.4 = 5.0) was not unusual compared to the other states (shaded gray in middle map). The right-hand map of the second row informs us that California's rate change from 1985 to 1990 (35.8 − 33.4 = 2.4) was less than that of most other states (shaded blue-green). This is more information than we had by looking at the usual rate maps alone. Furthermore, looking at the third row tells us that the rates were decelerating from 1985 to 1990; i.e., although the rates continued to rise, they rose less from 1985 to 1990 (2.4) than from 1980 to 1985 (5.0). This degree of deceleration was in the lowest 20% (fastest deceleration) of all the second-order changes (shaded green) over all states for all time periods. Again, this is more information than we could get from the usual rate maps.

We could speculate about why California female lung cancer mortality rates are changing in this way, which might lead us to hypotheses we could explore in other visualizations or test in epidemiologic field studies. Because of the complexity of lung cancer initiation, progression, diagnosis, and treatment, we can't say for sure what caused these trends and geographic patterns, but the changes in California noticed in this visualization are consistent with the results of a hierarchical statistical regression model that found that younger women in California had lower lung cancer mortality rates than older women, suggesting a cohort effect of declining cigarette smoking over time in that state (Pickle et al. 1996). The regression analysis and mapping of results took several days to complete, whereas this visualization was almost instantaneous once the generic code was prepared.

Perhaps a simpler explanation is to consider the state cancer priorities displayed in Figure 5.12, extending that cross-tabulation of states by level and trend of rates to include the changes in the trends, i.e., acceleration and

TABLE 6.2
A Conceptual Extension of Figure 5.12 to Include Acceleration and Deceleration for the Fifty-one States

Acceleration	Trend	Status		
		Above U.S.	Same as U.S.	Below U.S.
Accelerating	Rising			
	Stable			
	Falling			
Persistent	Rising			
	Stable			
	Falling			
Decelerating	Rising			
	Stable			
	Falling			

deceleration. A recent acceleration in regional cancer rates would raise concerns more so than a steady rising trend. The concept is illustrated in Table 6.2, which could also be displayed with micromaps in the table cells.

Applying physics concepts to the study of human disease may seem strange, but we hope you agree that this visualization method, although it takes some study to understand, did lead us to insights we would not have had from the original published maps.

6.6 ALTERNATIVE VIEWS

6.6.1 Temporal Change Maps with Alternative Views

Chunling Zhang and Dan Carr developed a Java implementation in 2006 called TCmaps (temporal change maps) that was a precursor to the comparative micromaps discussed in this chapter. Although the layouts have since been refined, we present some of the features of TCmaps here to show alternative views that might be of interest. TCmaps initially displayed three separate binary micromap sequences, each with its own class change map offset below (Figure 6.11). Each pair of binary micromap and class change map rows is framed in a light blue-gray rectangle. The top rectangle encompasses the high-value-class micromaps. The change map immediately below shows the states entering the top class in the next time period by shading with a solid color that indicates its previous class membership. The change map also outlines the states leaving the top class with the color of the class it is going to. For example, from the 1980-1984 map to the 1985–1989 map of the middle row of Figure 6.11, New Mexico changed to the

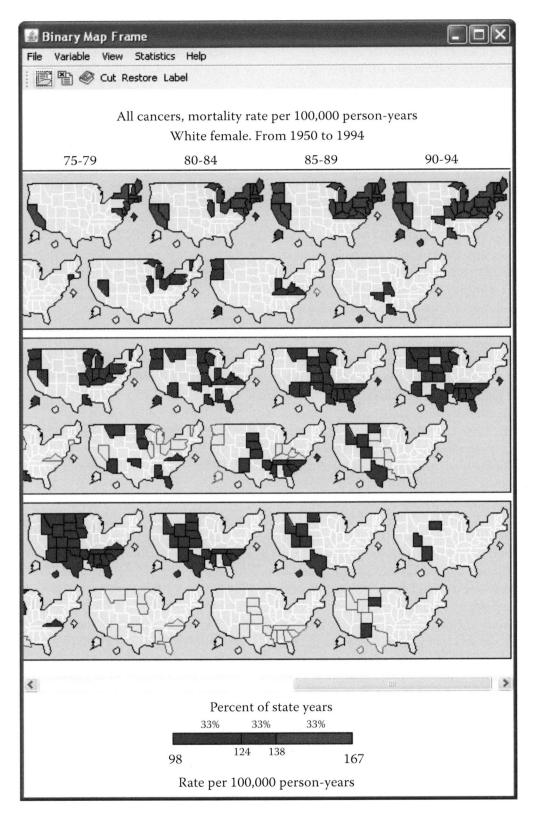

FIGURE 6.11   TCmap display of mortality rates for all cancer, white females, by state. Rising and falling rates are shown in separate pairs of rows.

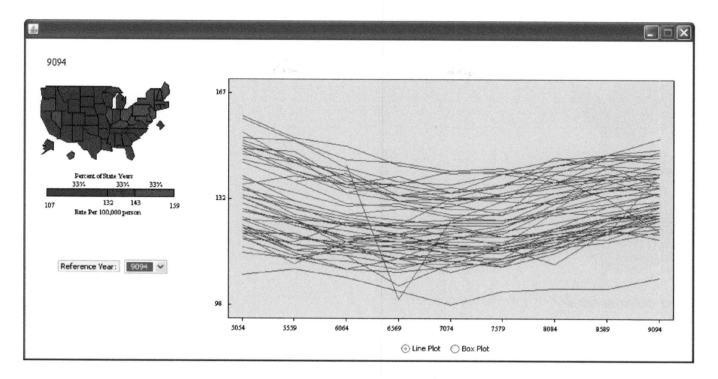

FIGURE 6.12 TCmaps display showing time period selection and the slider linked to both the map and the time series plot.

middle (magenta) category. In the change micromap below these two, New Mexico is shaded blue to indicate that it moved from the lowest (blue) category into the middle category. In the next change micromap, New Mexico is outlined in blue to indicate that it moved out of the middle category back to the lowest (blue) category for 1990–1994. Checking the third row, we can see that in fact New Mexico is blue in 1990–1994. The information about each class is complete in terms of displaying class change. The only local difficulty is that the outlined states are not as noticeable as the filled states.

The middle rectangle in Figure 6.11 encloses the middle class, with states in this category shown in magenta. This color had been chosen as a logical experimental choice because it is an additive color mix of red for the high class and blue for the low class (but we are not necessarily recommending this choice).

The horizontal scroll bar allowed interactive viewing of long time series. This version of TCmaps includes a vertical scroll bar, but this feature is not needed in Figure 6.11. However, the vertical scroll bar would probably still be needed in an application displaying large maps of many small regions, where screen size is limited so that all six rows could not be viewed at once. Alternatively, TCmaps has a simpler variant of a three row binary map design. This design drops the staggered change micromaps and color codes

each region in the three-class specific rows according to its previous class.

You can see the resemblance of this early TCmaps implementation to the new comparative micromap designs in this chapter. Color-coded micromaps, with categories defined by a user-controlled slider bar, paired with an offset row of change micromaps form the core of the new designs. However, the layouts in this chapter are more compact, eliminating the need to scroll vertically to see all of the rows.

In the alternative view shown in Figure 6.12, TCmaps provides a menu below the slider to select a time period for a map. The dynamic slider controls both the color of the states in the map and the color of the states' time series lines linking them together. Another available option replaces the time series line plot with a box plot view for each year.

Figure 6.13 shows a TCmaps alternative view with many options. This view resembles linked micromaps, but each of the micromaps on the left has its own dynamic three-class slider. The top two maps are from adjacent time periods. There is a scrolling slider below the arrow plots that selects the pair of time periods. The very bright third micromap shows the difference between states' values for the two time periods and provides a three-class look at differences, as opposed to class changes.

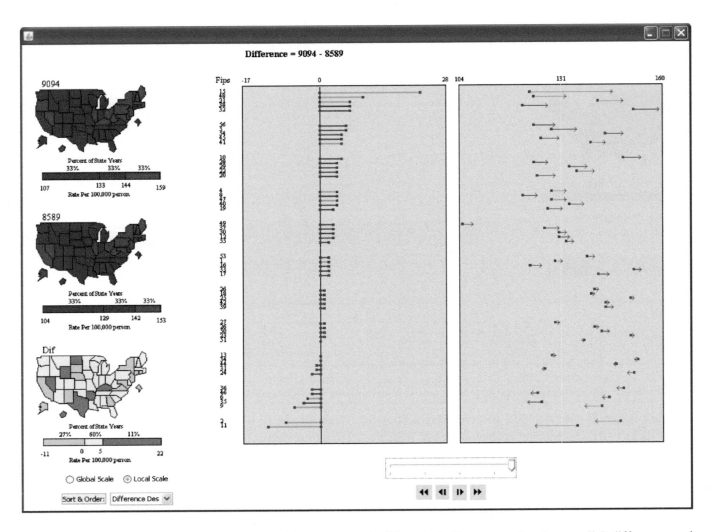

FIGURE 6.13   A feature-rich TCmaps display with three dynamic sliders, including one used to show explicit differences and statistical plots to sort and display the differences in several ways.

The use of arrows in the rightmost panel is similar to linked micromaps except for the lack of perceptual grouping lines. The arrow length shows the magnitude and the arrowhead shows the direction of change. The line with dots on the leftmost panel subtracts the base value from the first time period's value. This allows rescaling of the plot and calls even more attention to the direction and magnitude of change.

The buttons at the bottom, below the maps, provide the choice between global and local scaling. The sorting menu provides eight different sorting criteria for the arrow plots. Of the alternative views available in TCmaps, this one is the richest in features. TCmaps software is available on our website.

6.6.2   MORE GENERAL GEOVISUALIZATION TOOLS

The comparative micromaps described in this chapter, both the current version and its precursor TCmaps,

were designed for a fairly specific task, i.e., comparing geographic pattern changes over time. Other designers have taken a different approach and have included some tools for this task in more general tool kits. We describe a few of the more prominent ones here for comparison to our approach.

Alan MacEachren's group at Penn State has done extensive research on geovisualization for many years. Working with Linda Pickle at the National Center for Health Statistics, they developed a system called HealthVisB to facilitate spatiotemporal analysis (MacEachren et al. 1998). This prototype was used to test the effectiveness of animation versus stepping through single maps, finding that some people preferred one method over the other, but that some geographic changes could be missed with either method.

Working again with Linda later at the National Cancer Institute (NCI), they developed a Java application called ESTAT (Exploratory Spatio-Temporal Analysis

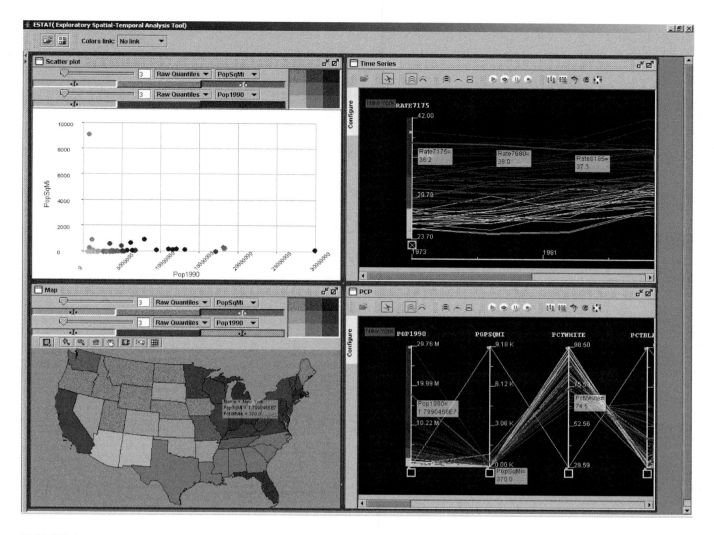

FIGURE 6.14 Exploratory Spatio-Temporal Analysis Toolkit (ESTAT) developed by Alan MacEachren's GeoVISTA Center at Penn State.

Toolkit) to provide NCI analysts a means to examine changing spatial patterns and associations between cancer rates and covariates of interest (MacEachren et al. 2003b). This more general data exploration program linked windows with a map, a scatterplot, a time series line plot, and a parallel coordinates plot by color, allowing the user to select the data points from any window. An early version is shown in Figure 6.14. Extensive usability studies with both students and NCI epidemiologists led to improved data import capability and additional statistical functionality (Robinson et al. 2005). The system was rewritten on the GeoVISTA Studio platform to provide more flexibility of design by linking Java beans appropriate for a particular application. One implementation integrates tools for multivariate data exploration, including ordering, sorting, bivariate views, and small multiples of maps and graphs (MacEachren et al. 2003a).

Another iteration of software development led to Improvise, a graphical user interface built on top of a modular library of data visualization components that are combined for each specific application. An interesting example is given in Figure 6.15. Based on guest registers from several historic hotels in Pennsylvania, the spreadsheet-like window on the left has days numbered in columns with a slider that allows the user to control the number of days to be displayed (Weaver et al. 2007). The spreadsheet cells and locations on the map are color coded according to the number of visits on a particular date (calendar) or from a particular hometown (map). One purpose of this analysis was to look for repeat visitors and identify any cyclic pattern in their visits. By changing the width of this sliding time view from the usual seven-day window to a fourteen-day window the analyst identified regular biweekly visits, most likely by traveling salespersons.

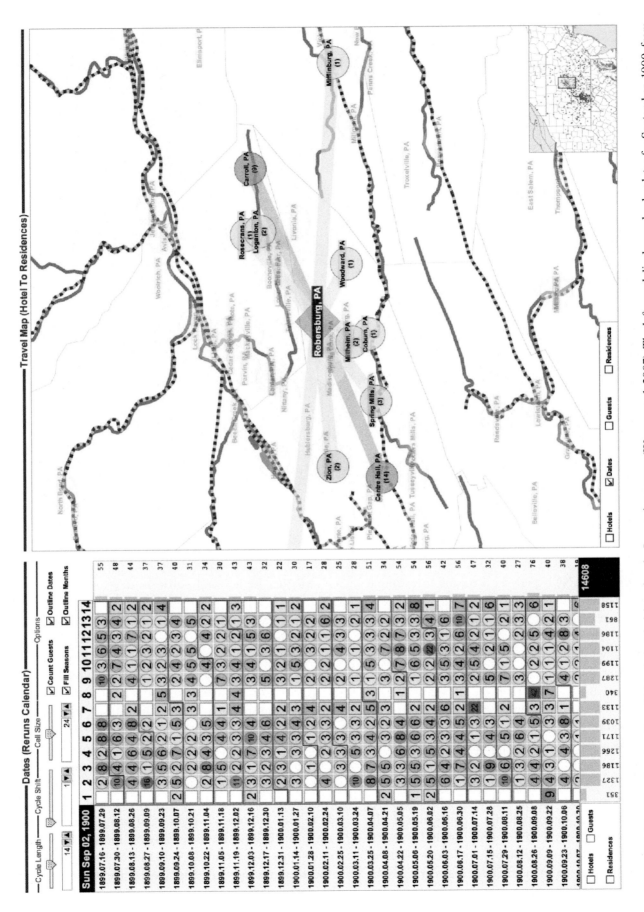

FIGURE 6.15 Visual exploration of historic hotel visits using the Improvise system (Weaver et al. 2007). The left panel displays calendar dates for September, 1900, from top to bottom. A slider controls the number of days shown in each row, here fourteen. The number of visits is encoded by color. This temporal display is linked to the map by color, showing the home towns of hotel visitors. (Image courtesy of Chris Weaver, University of Oklahoma, School of Computer Science and Center for Spatial Analysis.)

A popular data visualization program has been XGobi, with the new version renamed GGobi, by Swayne and colleagues (2003), and has now been linked to the R programming language (Ihaka and Gentleman 1996; Cook and Swayne 2007; R Development Core Team 2009). This software includes many of the modern-day visualization tools, but was not specifically developed for geographic data, except point data that could be viewed by using its generic scatterplot tools. The link to R provides a host of other tools, including some for spatial analysis. In contrast, another popular program, GeoDa by Luc Anselin (Anselin et al. 2006), has many tools for spatial data analysis but lacks some of the functions and flexibility for extensive exploratory visual examination of the data.

6.7 SUMMARY AND FUTURE DIRECTIONS

In this chapter, we have presented prototypes of comparative micromaps, a sequence of maps to be compared over some index, usually time. Of course, the values mapped over time must be comparable; a single slider defines a common categorization of these values that determines the color assigned to each of the micromaps. The simple two- and three-row layout may be expanded to a two-way design so that comparisons can be made simultaneously across the time periods as well as between two or more subsets of the data, e.g., males and females. Another extension demonstrated was the addition of another row of change micromaps representing the change in trends. Although these designs are fairly new and have not been subjected to specific usability tests, they are based on sound cognitive principles, such as the simultaneous display of small multiples of graphics for comparison, and design components proven effective in linked or conditioned choropleth micromaps. Within the next few years, we expect that the comparative micromap designs will be refined and mature software will be made available, but in the meantime, we hope that we have given you ideas that might be useful for your own data and applications.

7 Putting It All Together

7.1 SUMMARY

We hope you have enjoyed this tour of micromaps and now have a vision of how you can put micromaps to work in exploring data and communicating with others. Micromaps organize and present statistical data in static, interactive, and dynamic formats. All of the micromap designs stress comparisons for exploration and integrate statistical information with the geographical context. By basing the design elements on the current understanding of our cognitive strengths and limitations, we remove some of the cognitive burden required to visually process and understand the micromap. This makes the designs easy to use. Also, the statistics included in micromaps are familiar to many people without an extensive statistical background, so most people should be able to use these designs.

The theme of comparison and interpretation enabled by similarity and differences underlies much human activity, including scientific inquiry. Micromap comparisons can be spatial, temporal, between attributes, with an internal comparison, relationships among attributes, and at different scales. Successful designs usually are simple in appearance and present information in a concrete and credible manner. They also attract the reader immediately by telling a story or by displaying something unexpected in a familiar context, leading to speculation about the causes of noted patterns.

If what we first see is inconsistent with our personal worldview (cognitive dissonance), we may avoid looking at the graphic, and even if we do look, we may be unable to see or comprehend it. This is analogous to our vision—if the images seen by each eye are too different, we cannot fuse them to see in depth. Both similarities and modest differences enable us to see in depth. Consequently, it is often helpful for at least the initial micromap view to be of familiar geographic boundaries or displayed from an egocentric view. Although we have an amazing ability to shift perspective, we are not always able to see the world as other people do.

Linked micromaps link statistical graphics and small maps by color and by rows in a tabular format. Connection of the two components provides more information than either the graphic or the map alone, giving a geographic context to the statistics. This design has been used successfully to explore data interactively, such as by sorting the rows or selecting variables to display, and also in static form to communicate data or analytic results. This is the most mature of the three basic designs, having undergone extensive usability testing at the National Cancer Institute.

Conditioned choropleth micromaps facilitate thinking about three variables at a time in a spatial context: the study variable, such as disease rates, and two conditioning variables. The design partitions a single choropleth map into subset panels where regions are highlighted only if they meet the criteria set by the row and column slider bars for the two conditioning variables. This is more of an interactive, analytic tool than linked micromaps. Some statistical tools are built in to aid in setting the slider bars, and views of the data in several graphical forms supplement the map view.

Comparative micromaps display a sequence of indexed maps on a single page for the purpose of comparing the geographic patterns over the index values. This is the newest type of micromap and has not yet been extensively used or tested, so the designs presented here should be considered prototypes or ideas for you to try on your own data. However, the designs are based on sound cognitive principles and therefore are expected to work well. For example, our brains have limited capacity to attend to and remember multiple complex images such as a map. Therefore, the fundamental comparative micromap design is to display maps over the index simultaneously, e.g., on a single page, and to show in a different row the explicit differences between each pair of indexed maps. These design elements remove the cognitive burden of remembering multiple complex images and mentally subtracting values between maps, so that the reader can focus on examining the geographic patterns on the original maps or the difference maps. The design also alleviates the change-blindness problem and often makes it unnecessary to look back and forth between maps to find changes. We have shown that this design can be extended to work over two indices, such as time and gender.

Although there is some overlap in the type of comparisons each micromap style can facilitate, they generally serve different purposes and thus can be used as a set of tools for the exploration of geospatial data. To illustrate, we next present an exploration of Louisiana population data before and after Hurricanes Katrina and Rita in 2005. This is not meant to be an exhaustive analysis of these data but can serve as an exemplar for your own data exploration.

7.2 EXPLORATION OF LOUISIANA POPULATION CHANGES AFTER THE 2005 HURRICANES

7.2.1 BACKGROUND

The 2005 hurricane season was the worst in history, resulting in over two thousand deaths and over $128 billion in damages (Wikipedia 2009). Of the seven major hurricanes (Category 3+), two of them passed over Louisiana. Hurricane Katrina was a huge storm that slammed into New Orleans on August 29, 2005. Not only was there extensive damage in southeastern parishes (county equivalents) from the storm surge, but levees separating Lake Pontchartrain from New Orleans broke down, flooding about 80% of the city. Less than a month later, Hurricane Rita roared across the Gulf of Mexico and devastated parishes near the Texas border. Not surprisingly, regions of the state that had the deepest floodwater had more evacuees and repopulation has taken longer (McCarthy et al. 2006). While natural disasters strike populations randomly, Louisiana residents who were physically or financially unable to evacuate prior to the storm's landfall, such as the poor and the elderly, were the hardest hit and, some believe, received less recovery aid (Cohen 2008). The slow government response in distributing rebuilding funds further hampered recovery and return to the region by those unable to repair or rebuild on their own, leading to a potential shift in the demographic profile of the region (ACORN Housing/University Partnership 2007).

Most of you are familiar with this story, and so we will use Louisiana population data by parish to demonstrate how all three types of micromaps could be applied to the same data set in an exploratory visual analysis. Population data were obtained from the Census Bureau's annual population estimates program for 2000–2007, with supplemental counts by race/ethnicity from the Census' special studies in the Gulf Coast states hardest hit by the hurricanes (Koerber 2006). The estimates for 2005 were for July 1, 2005, prior to the hurricanes. Preliminary statistics showed that people who moved out of the region were more likely black, unemployed,

living in poverty, or living in rented housing (Koerber 2006). We will explore the changing population of Louisiana by some of these racial and socioeconomic characteristics. For illustration, we will use the parish as the basic geographic unit, recognizing that a more detailed analysis by census tract or other neighborhood definition would better represent the local changes in the region.

7.2.2 POPULATION CHARACTERISTICS BY PARISH

We start with a characterization of the Louisiana population prior to the hurricanes using data from the 2000 Census (Figure 7.1). The city of New Orleans is coincident with Orleans Parish. The Greater New Orleans region includes Orleans, Jefferson, St. Bernard, and Plaquemines parishes; the Census Bureau's Metropolitan Statistical Area definition adds St. Tammany, St. Charles, and St. John the Baptist Parishes. Lake Pontchartrain, whose burst levees flooded the city, lies between Orleans and St. Tammany Parishes. As is clear from Figure 7.1, much of the state's population is concentrated in this major metropolitan area. The racial makeup of the state varies by parish—the percent black, for example, ranges from less than 4% to over 67%. The percent Hispanic varies little across the state, with a maximum of 7%, and so is not included in this plot.

Re-sorting this plot by the percent black population (Figure 7.2), we can see that the median parish has about 30% black residents. Furthermore, parishes with a higher proportion of blacks than the median nearly form a band from New Orleans westward and northward, following the inland state borders (see cumulative micromap just above the median panel). So, while there is not a clear east–west or north–south delineation of parishes by race, there does seem to be a tendency for a lower proportion of blacks in southern coastal and western parishes. Hurricane Rita affected Cameron Parish, with a very low percent black, while Hurricane Katrina affected the Greater New Orleans region, which had a higher percent of blacks.

In looking at Figure 7.2, we thought we saw a tendency for parishes with a high percent black to also have a high percent of people living below the poverty line. The calculated correlation is 69%, so our eyes were not fooled. However, re-sorting again by the percent of people in poverty (Figure 7.3), we can see that the geographic clustering of high-poverty parishes is somewhat different than the clustering of high-percent black parishes—more of the high-poverty parishes are in the

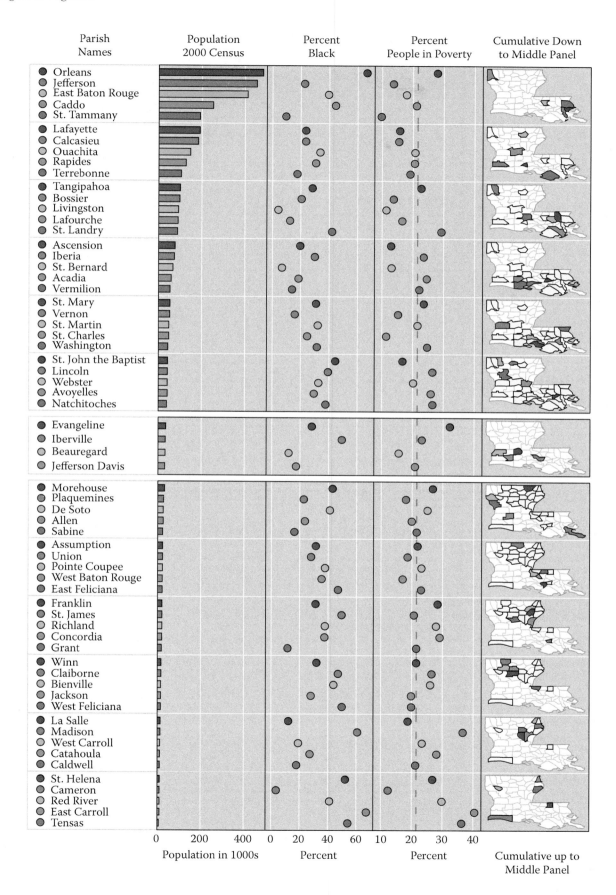

FIGURE 7.1 Characteristics of Louisiana parishes in 2000, sorted by population.

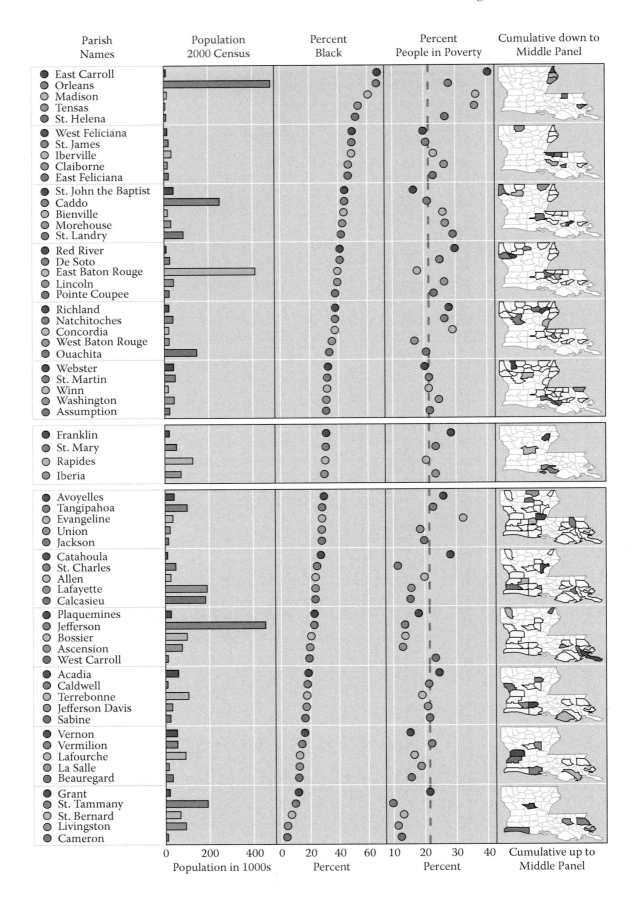

FIGURE 7.2   Characteristics of Louisiana parishes in 2000, sorted by percent black population.

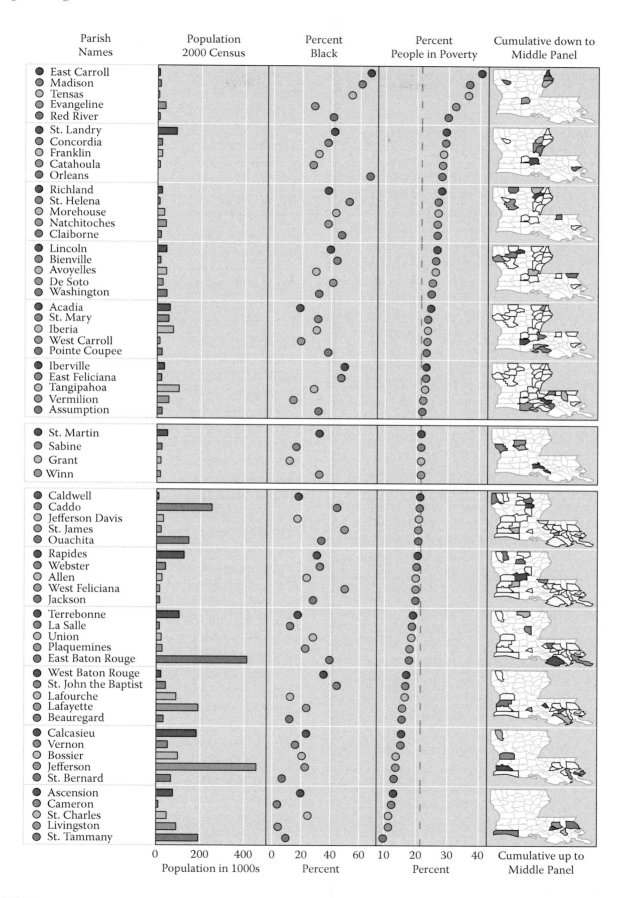

FIGURE 7.3 Characteristics of Louisiana parishes in 2000, sorted by percent of people living below the federal poverty level.

south-central portion of the state (see cumulative micromap just above the median panel). Parishes affected by the hurricanes had poverty levels below the median, i.e., were not as poor as other parishes.

7.2.3 Design Issues with the Louisiana Linked Micromap Plot

One design issue that is obvious from these three figures is the space constraint on the number of rows, here the number of parishes, that can easily be displayed on a single page. The maps have been vertically compressed in order to fit thirteen maps on the page. The font size of the row labels and the space between the rows have also been reduced. As micromaps get smaller, color becomes harder to identify for small regions. What had been easy to see can become difficult or impossible to see. Using wider panels to increase the area of the map regions displayed can help a little, but too much departure from the aspect ratio of familiar maps is disconcerting. Using larger perceptual groups to get by with fewer maps is undesirable since five per group is already beyond the guideline of four per perceptual group. Several of the parish names are rather long, and so the names column is wider than usual. Using word wrap to a second line for the longer names is not an option because that would impact the row height and spacing between the rows. Because there is an even number of rows (sixty-four), we do not separate out a single median parish, but use perceptual groups of size five with the extra four parishes shown in the center panels, surrounded by extra white space so that this group stands out. This is a hierarchical symmetric grouping of |5-5-5-5-5-5|4|5-5-5-5-5-5|, as discussed in Section 4.2.1 of Chapter 4. These space constraints are not a problem for dynamic applications where we can scroll down to see additional parishes. However, the disadvantage of scrolling is that we no longer can see all of the parishes at the same time.

The option of using a two-column format as illustrated earlier in this book can work but imposes a width constraint. Two columns of long names alone can take up a lot of the available width. The double-column format can usually work with a single statistics column. However, we wanted to show several statistics columns, so we proceeded with smaller fonts and maps.

7.2.4 Population Changes after the 2005 Hurricanes

In Figure 7.4 we can see where the worst hurricane flooding occurred. St. Bernard, Orleans, and Plaquemines Parishes lost 78, 44, and 22% of their population, respectively, in the six months after Hurricane Katrina (2005 to 2006). Cameron Parish in the southwestern corner of the state lost 19% after Hurricane Rita hit about a month later. The devastation in these four parishes dwarfs population changes in the other regions, so in the remaining linked micromap plots we will separate these four out from the others. Comparing the initial changes in population to those during the following year (2006 to 2007), the two hardest-hit parishes started to recover, but the next several parishes continued to lose population (note that we have added a reference line at 0% change to aid in identifying gains versus losses in these two columns).

This is clearer in Figure 7.5, which shows percent changes in population by an arrow. The base of the arrow represents the percent change in population from 2005 to 2006, and the arrow tip represents the cumulative percent change from 2005 to 2007. It is important to note that the denominator of both percentages in this figure is the 2005 population so that the percentages are comparable. In contrast, the denominator of the 2005 to 2006 change column in Figure 7.4 is the 2005 population, but the denominator of the 2006 to 2007 change column is the 2006 population. It may seem confusing at first that the percent calculations changed between these two figures, but otherwise the arrow symbology could not have been used.

St. Bernard Parish gained a remarkable 39% relative to its 2006 population, but it has a long way to go before returning to its pre-Katrina size (Figure 7.5). The bar plot showing the 2005 population provides this context. The percent change in St. Bernard Parish is big, but in terms of people not as big as the increase in Orleans Parish. We can see by the direction of the arrows that Plaquemines and Cameron Parishes lost even more residents from 2006 to 2007.

As we have said several times throughout this book, in most cases we wish to compare values across all of our geographic regions on the same graphic. However, splitting out the four most affected parishes into a separate plot (Figures 7.5 and 7.6) allows us to use a different scale for the two plots based on the range of each set of parish values. In addition, we gain some vertical space by reducing the number of parishes slightly, and so the plot is not quite as crowded. This results in larger maps and larger regions, so that the colors are easier to identify, especially in small regions.

A question that might naturally arise at this point is where did the people go? Many were evacuated to Houston and other cities that had the facilities and social

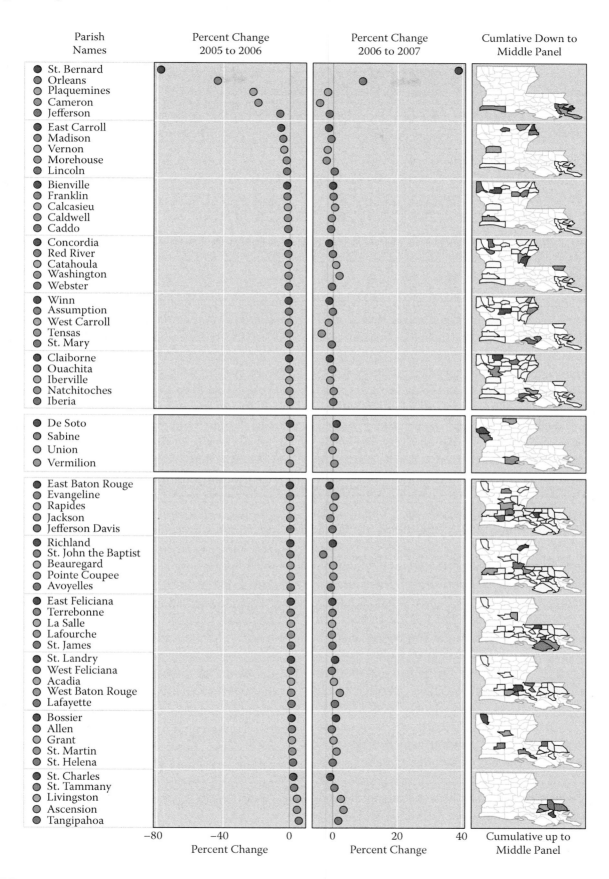

FIGURE 7.4 Percent change in each parish's population after hurricanes hit in August/September 2005, and the recovery in the next year (2006 to 2007).

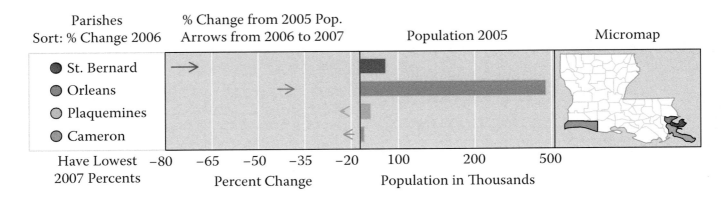

FIGURE 7.5 Cumulative population change from 2005 to 2007 for four parishes that were the hardest hit by 2005 hurricanes. Base of arrow represents the percent change in population from 2005 to 2006 (sorting variable), and the arrow tip represents the percent change from 2005 to 2007.

services to best help them. The Current Population Survey, jointly conducted by the Census Bureau and the Bureau of Labor Statistics, included questions for hurricane evacuees during 2005 and 2006 to determine the demographic makeup and movement of these people (Groen and Polivka 2008). Some of these out-of-state evacuees never returned to the parish where they lived at the time of the hurricane, but many residents returned to their home parish after a temporary stay elsewhere. Can we draw any conclusions from a simple visualization of population changes?

In Figure 7.7, we examine the changing populations within the state from 2005 to 2006 by a conditioned micromap plot. All three sliders' thresholds are set to changes of –1% and +1% (or as close as possible, given ties and the pixel width of the vertical lines defining the slider thresholds). Color is determined by the percent change in non-Hispanic whites (hereafter shortened to "whites"), so parishes shaded red had an increase in their white population of at least 1%. The percent change in the Hispanic populations is defined by the rows—parishes highlighted in the top row gained and those in the bottom row lost Hispanic residents. Similarly, the percent change in the black populations is defined by the columns—parishes highlighted in the left column lost and those in the right column gained black residents. This plot has most of its shading in the lower left and upper right panels. The lower left panel shows the loss of the white population by blue shading in the four hurricane-affected parishes plus Vernon Parish, while the position of the panel tells us that these five parishes also lost blacks and Hispanics. Similarly, all of the red shading (gains in the white population) is in the upper right panel, indicating that there also were increases in the black and Hispanic population in these places.

It appears from this plot that all residents were affected by the storms regardless of race or ethnicity. About half of the parishes that gained white residents during this year were to the northwest of the Greater New Orleans region. This suggests that the white evacuees who stayed within the state did not move too far from home. Black evacuees seemed to move into many of the parishes (those highlighted in any color in the top right panel), with their destinations less concentrated around New Orleans than whites.

In Figure 7.8, the same conditioned micromap design is applied to the changes from 2006 to 2007, a time when residents returned to their homes if they could. Nearly all of the parishes are highlighted in the top row and none are highlighted in the bottom row. This indicates an increase of 1% or more of the Hispanic population across most of the state; no parish had a declining Hispanic population. Most of the red highlighting is in the upper right panel, suggesting that all three groups returned to St. Bernard Parish, one of the hardest hit, and other parishes north and west of New Orleans. There was at least a 1% increase in the black population of all of the highlighted parishes in this panel, even though the white population continued to decline in Cameron Parish (southwestern parish) and three parishes south of New Orleans. These slider settings were chosen to be consistent with those for the 2005 to 2006 changes in Figure 7.7, but variation in the white population change is not well explained here (R^2 is only 8.2%).

7.2.5 CHANGES IN POPULATION BY RACE AND POVERTY LEVEL

One of our early hypotheses was that there might have been a disparity by income, if not in the initial

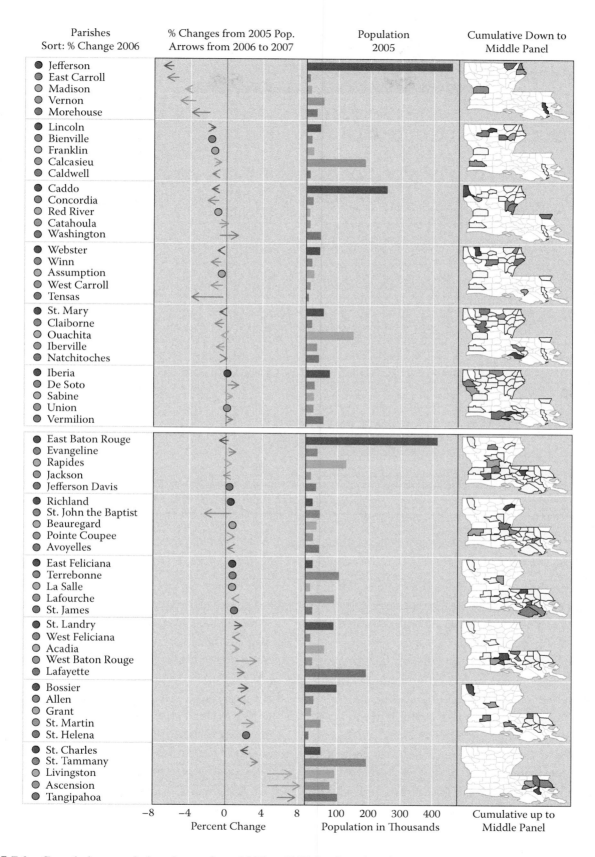

FIGURE 7.6 Cumulative population change from 2005 to 2007 for the other sixty Louisiana parishes. Base of arrow represents the percent change in population from 2005 to 2006 (sorting variable), and the arrow tip represents the percent change from 2005 to 2007.

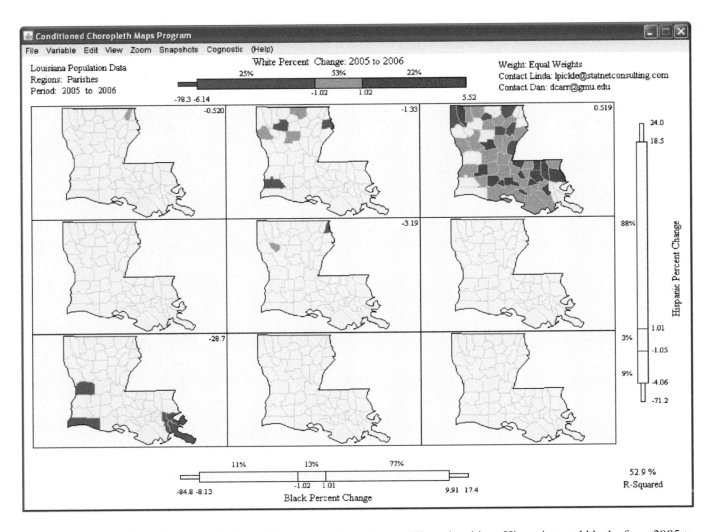

FIGURE 7.7   Conditioned micromap plot of the percent change in non-Hispanic whites, Hispanics, and blacks from 2005 to 2006 by parish in Louisiana.

displacement, then in the likelihood of return. The next figure (Figure 7.9) examines the population changes over the two years simultaneously, along with a stratification of parishes by poverty level. The slider settings were determined by the cognostic, which searched for a high R^2 value (95%). The few parishes highlighted in the left column lost the greatest percent of the white population from 2005 to 2006, but the red highlighting indicates that these also had large increases the next year as residents returned. For the entire state, the white population declined 3.3% from 2005 to 2007. The three parishes that gained white population from 2005 to 2006 continued to gain during the following year. By comparison, the level of poverty does not help to predict the percent change from 2006 to 2007 once we know the earlier population change. Comparing the row means within each column does not reveal much of the conjectured avoidance of high-poverty areas. Adjusting the color

slider changes the appearance of the micromaps but not the within-panel statistics.

Figure 7.10 repeats the conditioned micromap design for blacks, with the color slider thresholds set to evenly divide the parishes among the three colors. The few parishes highlighted in the left column show that many blacks returned to places that had seen large losses in 2005. Those highlighted red in the right column had an increase in the black population following an increase in the earlier year; i.e., they continued moving to these parishes. These patterns are similar to those of whites (Figure 7.9), but fewer blacks had returned to the state by 2007 (state population decline of 7.5% for blacks, 3.3% for whites). In contrast to whites, poverty appears to be a strong predictor of the percent change in the black population. All of the blue regions, parishes that lost blacks during 2006–2007, appear in the top row that is associated with higher levels of poverty. The

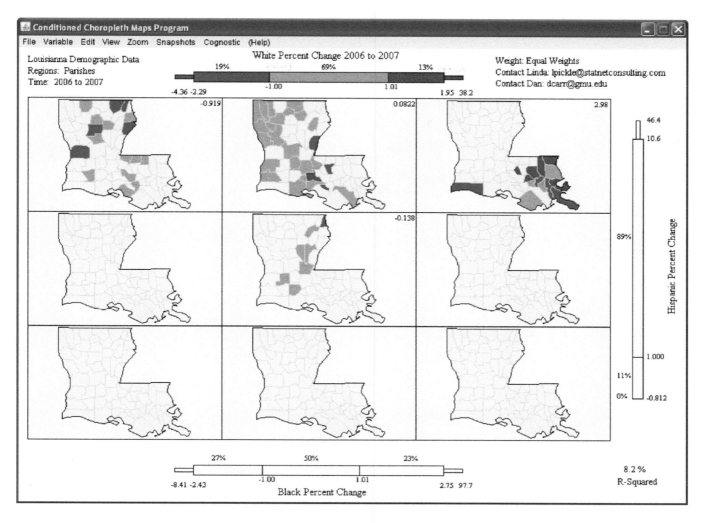

FIGURE 7.8 Conditioned micromap plot of the percent change in non-Hispanic whites, Hispanics, and blacks from 2006 to 2007 by parish in Louisiana.

gradients of the means in the middle and right columns are consistent with poverty avoidance; i.e., blacks seemed to be returning to parishes that had fewer other residents living in poverty.

Looking at the CCmaps scatterplot smooth view did not provide any evidence of a poverty effect for whites. A graph of the percent changes from 2006 to 2007 by poverty as a continuous variable (right plot of Figure 7.11), eliminating the categorization of poverty and the display of the earlier population changes, still does not show any association among whites—the scatterplot smooth is essentially flat. The population changes also seem equally likely to be increases as decreases (parish values seem evenly scattered about the zero line).

On the other hand, the scatterplot smooth view of the black data led us to an "aha." What seemed to be a pattern in the panels with the low-, middle-, and high-poverty encodings became clearer when viewing both variables as continuous. The population increases from 2006 to 2007 were most often associated with parishes that had low percent poverty in 2000. The left plot of Figure 7.11 gets to the essence of what we saw by omitting the black population percent changes from 2005 to 2006 and values for three outlier parishes. The scatterplot smooth clearly shows that parish populations increased more in parishes that had had lower poverty levels in the 2000 Census.

7.2.6 LONGER-TERM POPULATION TRENDS BY RACE

Our examination of the population changes by parish in Louisiana has been limited to three years, 2005 to 2007. In part, we wanted to focus on the time immediately following the hurricanes, but the time span was also limited by the designs of the linked and conditioned micromaps. Time series can be shown in linked micromap plots (see Figure 4.14), but these can be difficult

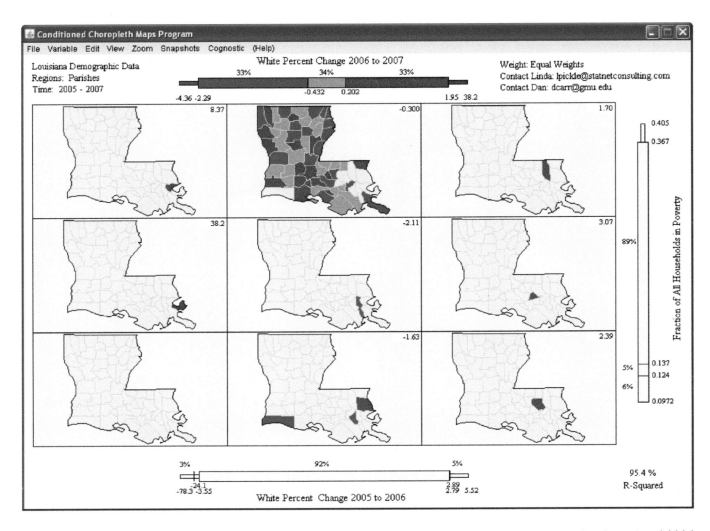

FIGURE 7.9 Conditioned micromap plot of percent change in the white population from 2005 to 2006 (columns) and 2006 to 2007 (highlighted) by poverty level in the parish (rows).

to read unless there are clear time trends. Arrows can be used, as we did, to show change from one year to another but, without additional micromaps, are not suitable for showing changing spatial patterns in longer time series. Figures 7.12 to 7.14 display the longer time series of population and changes for whites, Hispanics, and blacks, respectively, using the comparative micromap design. Each of the main maps in the center row shows the percent change in that population from one year to the next, with the color slider thresholds set at –1% and +1% as before. Note that although exact percent changes per year are not shown, any fairly constant increase of even 1% per year is characteristic of exponential population growth.

Figure 7.12 indicates that whites had been leaving northern Louisiana parishes and moving to regions around New Orleans prior to the hurricane. Plaquemines Parish (southeastern parish) had been gaining white residents prior to 2005 but had a declining white

population after that time. Figure 7.13 shows the corresponding population change data for Hispanics. Unlike for whites, now every map looks the same. Except for a loss in Hispanic population immediately following the hurricane, the Hispanic population in nearly every parish has been growing more than 1% per year over this entire time span. Figure 7.14 confirms our finding that blacks displaced by the hurricane moved to nearly every other parish in the state, but we can also see that the black population had been increasing in eastern parishes near New Orleans prior to 2005 and continued to do so after 2006.

7.2.7 SUMMARY OF THE ANALYSIS

This has been an illustrative, not an exhaustive, analysis of population changes in Louisiana after the 2005 hurricanes. Recalling our STAIRS acronym, we examined the data from multiple points of view:

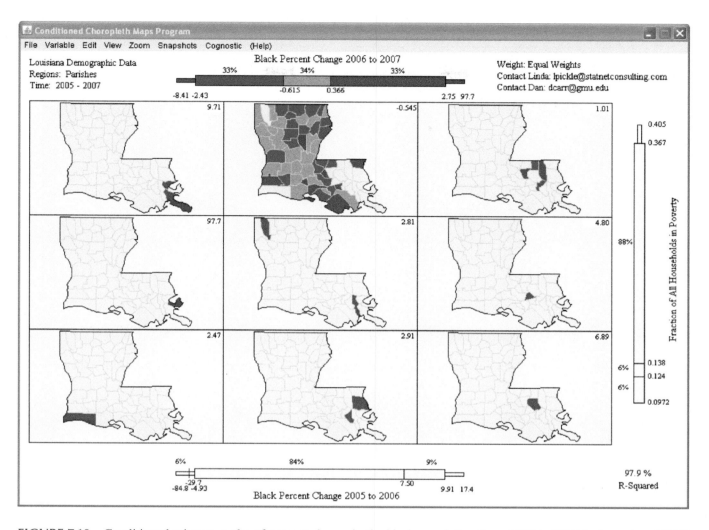

FIGURE 7.10 Conditioned micromap plot of percent change in the black population from 2005 to 2006 (columns) and 2006 to 2007 (highlighted) by poverty level in the parish (rows).

Spatial—Geographic patterns of percent change.

Temporal—Changing geographic patterns over time.

Attribute—Compared map patterns and parish ranks by poverty levels and racial makeup.

Internal comparisons—Compared parishes affected by the hurricanes to others in the state.

Relationships among the variables—Compared the rank order of parishes by percent blacks and poverty level.

Scale—The analysis was primarily an examination of parish populations, although we did note population changes in the Greater New Orleans region and for the whole state over several years.

We first included parish populations in the linked micromap views to give a context for percent change, but the later micromaps examined percent change as a way to standardize the study variable for the wide variation of population sizes by parish. We set the row and column slider values near –1% and +1% in the conditioned micromap views to help us think about the population changes in a consistent way across race and time using easy-to-remember thresholds. Setting the thresholds to larger changes will display somewhat different patterns, but we thought that this setting would show most of the substantial changes without being too affected by random year-to-year variation (percent change near 0). Because we were primarily interested in comparisons of population changes over parishes, the study variable for this analysis is annual percent change (2005–2006 and 2006–2007). This approach ignores underlying differences in population counts, i.e., is unweighted by population size. An analyst who is more interested in measuring the burden on parish-specific social services, for example, would prefer to use actual population counts as the study variable.

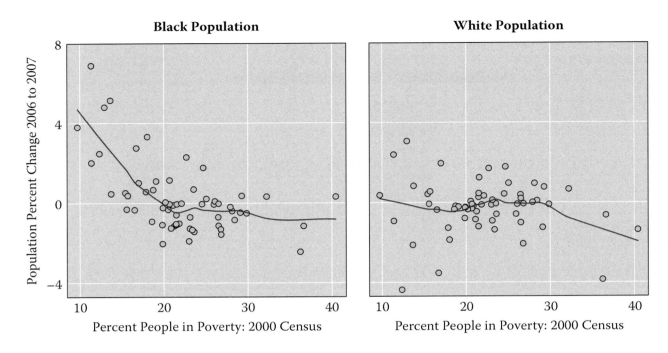

FIGURE 7.11   Plot of percent change in the black and white populations from 2006 to 2007 by poverty level in the parish.

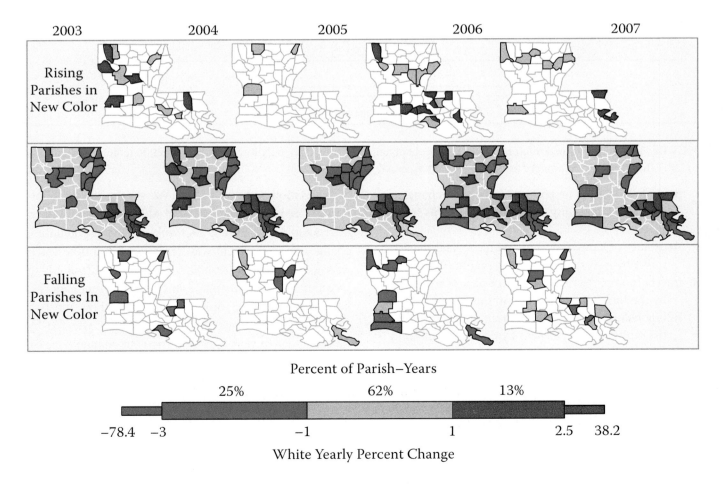

FIGURE 7.12   Annual percent change in white populations by parish, 2003–2007, with parishes that changed categories shown in additional rows.

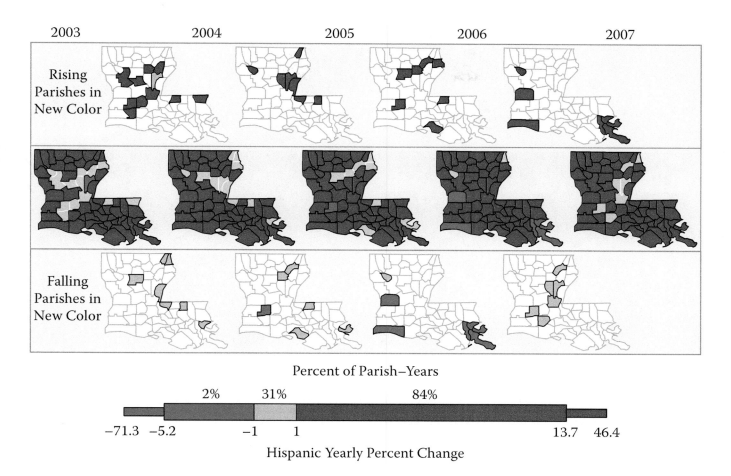

2003   2004   2005   2006   2007

Rising Parishes in New Color

Falling Parishes in New Color

Percent of Parish–Years

2%   31%   84%

−71.3 −5.2   −1   1   13.7   46.4

Hispanic Yearly Percent Change

FIGURE 7.13   Annual percent change in Hispanic populations by parish, 2003–2007, with parishes that changed categories shown in additional rows.

Our exploratory visual analysis has found changes consistent with other reports based on more complex analyses (Groen and Polivka 2008), but also a few that appear to be new. We found that populations declined in the affected parishes across categories of poverty, race, and ethnicity, consistent with the Current Population Survey. However, Hispanics and blacks evacuated to a wider region within the state than whites, who tended to move just beyond the affected parishes. (Of course, it is possible, but not probable, that the increases in populations were due to individuals moving from other states. We have parish population without information about where individuals moved.) In the year following the disaster, there were large increases in population for two of the most affected and most populous parishes, St. Bernard and Orleans, but two others, Plaquemines and Cameron, continued to lose residents. The population for all three race/ethnicity groups increased in parishes north and west of New Orleans during this time. Splitting the parishes by population changes in each year showed that both blacks and whites tended to

move to parishes in 2006–2007 that had the extremes of population change in 2005–2006, i.e., had both large losses and large gains in population in the first year following the hurricane. The choice of blacks (but not whites) to return to a particular region also appeared to be related to levels of poverty by parish—black populations increased in parishes with lower levels of poverty. Over a longer period of time, we saw that the Hispanic population was steadily increasing all across the state, but white and black populations tended to rise more in the New Orleans metropolitan area. It may be that the choice of where to live in the future was influenced by where the early evacuees finally settled as well as by potential job opportunities and social services.

7.3   CONCLUDING REMARKS

We have presented three types of micromap plots in this book, with many variations of each. What is best for your application will depend on a number of things, not the least of which is knowing the goals of

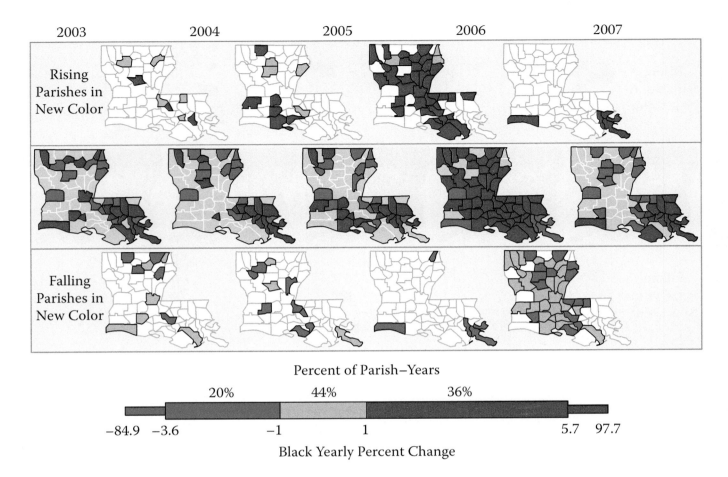

FIGURE 7.14 Annual percent change in black populations by parish, 2003–2007, with parishes that changed categories shown in additional rows.

the visualization and the skills of the targeted audience. Designing micromaps for communication of statistics to a wide audience will require more care than designing an exploratory micromap for personal use. Usability studies are especially important for designing effective communications, but can be equally helpful for designing interactive tools to be shared among analysts. We know that some of you will see things to improve in our designs. Perhaps there is even a whole new class of micromaps we have missed. We encourage such thinking and hope that results from cognitive science will be increasingly coordinated with the evolution of quantitative graphics.

Of course, it is not helpful if we present graphics of very precise statistics but few can understand or use the results. Some of the statistics used in micromap designs are means, percents, and box plots, which are now taught beginning in elementary school. Confidence intervals and hypothesis tests are typically presented in the first course in statistics. Since these are commonly used statistics, they should be familiar to many people. However,

one goal of micromaps is to provide educational pathways toward more sophisticated statistical methods. We have also attempted to provide insights into data analysis techniques that are learned only by experience.

There are some issues that still weigh on our minds. The power of graphics to aid understanding is well recognized, but with power comes the risk of misuse. Some people advocate the restriction of graphs and data to avoid misuse or to avoid drawing attention to problems. As educators we seek to provide both tools and education with the hope that learning will continue. Graphics can be misused, but our position is that people can learn from mistakes. We also believe that when many people can see and share perspectives, we are in a better position to see constructively and shape the world.

Access to data is a big issue. There are an increasing number of champions of open access, data standards, and shared tools. Progress is being made. In this book we have made heavy use of data provided by U.S. federal agencies, and there are often quiet heroes who some time ago fought key battles to make data available.

We call attention to just one champion from the international scene, Hans Rosling. He has not only pressed for data access to bring data to light, but he has also masterfully used graphics and story telling that bring data to life.

Access to regional boundary files and language are also barriers. Our future plans include examples for other nations along with boundary files and software for different languages. Our progress has been slow and a larger-scale solution is desirable.

There are a growing number of data visualization tools, such as Gapminder World and the three forms of micromaps shown here. We are no longer dependent on highly specialized analysts to process and interpret data for us. Of course, we can still benefit from insights of specialists, in particular those with knowledge of the subject matter and data collection methods who can help us to interpret what we see. Some caution is appropriate, though, because we can often see what we want to see. Still, sometimes data speak very clearly.

We also hope that developers of commercial software will begin to incorporate the micromap plot designs into their packages, to marry these visualization designs to their better computational power for data manipulation and modeling. The designs presented in this book may, of course, evolve over time as cognitive psychologists continue to learn about how we see and think. There are also interesting ideas in the past work of others that could be better integrated into new or improved tools, some that are simplifications of current designs.

Remember that micromaps are applicable to data from a wide range of disciplines. Our examples that lean toward health, environment, and demographics are due largely to data availability and our backgrounds. The mortality and population data are available for all U.S. states, and so could easily be used to demonstrate our designs. It is important to examine disease data such as cancer mortality to gain an understanding that might improve the situation. For example, cancer mortality rates have been declining for a few years, in no small part the result of early maps and epidemiologic studies that examined the reasons for so many cancer deaths in particular geographic regions ("hot spots;" Mason et al. 1975; Mason 1995). Some progress has been made via national overview.

However, many people see problem areas more locally, more directly, and are busy taking constructive action in their daily lives. Educators, health care workers, and farmers around the world are taking constructive action, seeking to make the world a better place for our children and grandchildren. In the service of this goal, there is room for perspectives, understanding, and action at all scales, from the very local to nationally and internationally. Graphics can help people to see through the noise to the patterns in the numbers and to learn from the lives of people to obtain a better perspective on the challenges being faced by real people. While quantitative graphics can help reveal patterns, they don't substitute for action. Nonetheless, graphics may motivate and guide action. We offer our simple graphics designs and tools in the hope that more people can study the numbers and use what they see to learn more and to take appropriate action.

One of our guiding principles throughout the book has been to engage the analyst. We all can do so much more if given tools to pursue leads provided by the mixture of data and our experience. We are inclined to do more when we can pursue our own direction of thought. In this book we suggest simple designs and tools, since these help to engage people and enable them to take next steps. While firm answers may require scholarly research, experiments, data collection, sophisticated modeling, and substantial community consensus, the early next steps often generate insights that motivate deeper inquiry. Many next steps can be faltering, but if we don't take next steps, we can't learn the joy of running. John Tukey's inspiration remains with us. He said it was okay to use what we see in the data to help guide us in building models. He enjoined us again and again to take next steps. He often reminded us to pause and see how far we have come. As we end this tour, now would be a good time to reflect on how far we have come in visualizing data patterns using micromaps.

Appendix 1: Data Sources and Notes

All URLs were operational on the access date. There should be enough information here to find the data in case the URL has changed recently.

1. Poverty and education level by state
   Source: Census 2000 Summary File 3, accessed through the American FactFinder website (http://factfinder.census.gov).
   Notes: "South" was defined to be the Southern Census Region.

2. Baseball statistics
   Source: Major League Baseball sortable player statistics for the 2007 season (http://mlb.mlb.com/stats/sortable_player_stats.jsp, accessed December 2, 2008).
   Notes: Data restricted to players with 50 at bats for hitting (n = 542 players) for slugging percent, 10 fielding chances (5 for pitchers) (n = 1,201 players).

3. Mammography use and health insurance coverage
   Source: CDC Behavioral Risk Factor Surveillance System (http://www.cdc.gov/brfss).
   Notes: We averaged the screening percents over 1998–2006 to get an annual percent. The mammogram question was asked in all states for 1998–2000 and thereafter only in even years. The health insurance question was asked annually and included any health care coverage, including HMOs and Medicare.

4. State unemployment by year
   Source: Bureau of Labor Statistics Local Area Unemployment Statistics Program (http://www.bls.gov/lau).
   Notes: Rates were averages of monthly rates over each year.

5. Mammal brain size
   Source: Chambers et al. (1983) appendix, a subset of a larger dataset of brain and body mass measurements from Allison and Cicchetti (1976).

   Notes: We assigned each mammal listed to a biological family according to the Diversity Web, University of Michigan Museum of Zoology (http://animaldiversity.ummz.umich.edu/site/index.html).

6. HIV mortality rates
   Source: National Center for Health Statistics; observed rate map published in the *Atlas of United States Mortality* (Pickle et al. 1996). Also available online at http://www.cdc.gov/nchs/products/other/atlas/atlas.htm
   Notes: Smoothed maps created by applying the Headbang algorithm (Mungiole, Pickle, and Simonson 1999).

7. Cancer mortality, 1995–2005 (lung cancer and all cancer)
   Source: Surveillance Research Program, National Cancer Institute SEER*Stat software (http://www.seer.cancer.gov/seerstat), November 2007 data submission, released April 2008; data originally provided to NCI by the National Center for Health Statistics (http://www.cdc.gov/nchs).
   Notes: Only malignant cancers were included.
   Unless otherwise noted, rates were directly age adjusted over nineteen age groups using the U.S. 2000 standard population. When these data were to be compared to 1950–1969 or 1970–1994 NCI data, these rates were directly age adjusted over nineteen age groups using the U.S. 1970 standard population for comparability.
   Corresponding populations were also obtained from the SEER*Stat database but were originally provided by the Census Bureau.

8. Lung cancer mortality, 1950–1969 and 1970–1994
   Source: National Cancer Institute Cancer Mortality Maps and Graphs website (http://www3.cancer.gov/atlasplus), data originally published in Devesa et al. (1999).

Notes: Rates were directly age adjusted to the U.S. 1970 standard population.

9. Pennsylvania toxic releases of air pollutants

Source: Environmental Protection Agency's Toxic Release Inventory data for Pennsylvania, 2007, available through the TRI Explorer (http://www.epa.gov/triexplorer/).

10. Primary syphilis rates in males by state, 1984–2003

Source: Centers for Disease Control and Prevention, National Center for HIV, STD and TB Prevention, available through the CDC WONDER website (http://wonder.cdc.gov/std-v2003.html).

11. Migration of residents to and from Iowa, 1995–2000

Source: Census 2000, Summary File 3.

12. Homicide rates among males, 2001–2005

Source: Surveillance Research Program, National Cancer Institute SEER*Stat software (http://www.seer.cancer.gov/seerstat), November 2007 data submission; data provided to NCI from the National Center for Health Statistics (http://www.cdc.gov/nchs).

Notes: The District of Columbia was excluded from the plots because its rate was a high outlier. We felt that DC was unlike the fifty states in that it is 100% urban, and so it was not comparable to the other states.

13. Biodiversity data—number of bird and mammal species

Source: Environmental Protection Agency, Western Ecology Division, Biodiversity Data website (http://www.epa.gov/wed/pages/staff/white/getbiod.htm).

14. World interest rate spreads, August 22, 2008, to February 5, 2009, in eight-week intervals.

Source: Financial Times government bond archive (http://markets.ft.com/ft/markets/researchArchive.asp?report=GOV).

Notes: The nineteen countries displayed had rate data available for these dates. Rates for the closest government bonds to two-year and ten-year maturities were used. The date in the graphic is the date for all or nearly all of the bond quotes; however, occasionally the quote was for a day before or after the displayed date.

15. Louisiana parish population data

Source: Census Bureau annual population estimates by parish were used for the long-term trends (2000–2007) (http://www.census.gov/popest). Data were accessed through the American FactFinder website (http://factfinder.census.gov/servlet/GCTTable?_bm=y&-context=gct&-ds_name=PEP_2007_EST&-mt_name=PEP_2007_EST_GCTT1_ST2&-CONTEXT=gct&-tree_id=807&-geo_id=04000US22&-format=ST-2|ST-2S|ST-2Sh&-_lang=en).

Notes: These estimates were for July 1 each year and included adjustments for the population changes due to Hurricanes Katrina and Rita. Estimates for 2005–2007 were based on the Census Bureau's American Community Survey (a rolling census count) Special Product for the Gulf Coast Area, which was conducted beginning in October 2005 (posthurricane). See http://www.census.gov/acs/www/Products/Profiles/gulf_coast/about.htm for further information. For our purposes, the 2005 population estimates were for just before the hurricanes, the 2006 estimates were for ten to eleven months posthurricane, and the 2007 estimates were about two years posthurricane.

Appendix 2: Suggested Symmetric Partitionings for Micromap Perceptual Groupings

#	Partitioning 1	Partitioning 2
1	1	
2	2	
3	3	
4	4	
5	5	
6	3 3	
7	3 1 3	2 3 2
8	4 4	
9	4 1 4	3 3 3
10	5 5	
11	5 1 5	3 5 3
12	5 2 5	4 4 4
13	5 3 5	4 5 4
14	5 4 5	
15	5 5 5	
16	5 3 3 5	4 4 4 4
17	5 3 1 3 5	3 4 3 4 3
18	5 4 4 5	4 5 5 4
19	5 4 1 4 5	4 4 3 4 4
20	5 5 5 5	
21	5 5 1 5 5	4 4 5 4 4
22	5 5 2 5 5	5 4 4 4 5
23	5 5 3 5 5	
24	5 5 4 5 5	
25	5 5 5 5 5	
26	5 5 3 3 5 5	5 4 4 4 4 5
27	5 5 3 1 3 5 5	4 4 4 3 4 4 4
28	5 5 4 4 5 5	4 5 5 5 5 4
29	5 5 4 1 4 5 5	4 4 4 5 4 4 4
30	5 5 5 5 5 5	
31	5 5 5 1 5 5 5	4 4 5 5 5 4 4
32	5 5 5 2 5 5 5	5 5 4 4 4 5 5
33	5 5 5 3 5 5 5	4 5 5 5 5 5 4
34	5 5 5 4 5 5 5	
35	5 5 5 5 5 5 5	
36	5 5 5 3 3 5 5 5	4 4 5 5 5 5 4 4
37	5 5 5 3 1 3 5 5 5	4 4 4 4 5 4 4 4 4
38	5 5 5 4 4 5 5 5	4 5 5 5 5 5 5 4
39	5 5 5 4 1 4 5 5 5	4 4 4 5 5 5 4 4 4
40	5 5 5 5 5 5 5 5	
41	5 5 5 5 1 5 5 5 5	4 4 5 5 5 5 5 4 4
42	5 5 5 5 2 5 5 5 5	5 5 5 4 4 4 5 5 5
43	5 5 5 5 3 5 5 5 5	4 5 5 5 5 5 5 5 4
44	5 5 5 5 4 5 5 5 5	
45	5 5 5 5 5 5 5 5 5	
46	5 5 5 5 3 3 5 5 5 5	4 4 5 5 5 5 5 5 4 4
47	5 5 5 5 3 1 3 5 5 5 5	4 4 4 4 5 5 5 4 4 4 4
48	5 5 5 5 4 4 5 5 5 5	4 5 5 5 5 5 5 5 5 4
49	5 5 5 5 4 1 4 5 5 5 5	4 4 4 5 5 5 5 5 4 4 4
50	5 5 5 5 5 5 5 5 5 5	
51	5 5 5 5 5 1 5 5 5 5 5	4 4 5 5 5 5 5 5 5 4 4

The left column (#) contains the number of regions. Partitioning 1 puts the smallest counts in the middle group. Partitioning 2 alternatives are also fully symmetric but avoid small counts. Abandoning full symmetry can lead to fewer panels. Only 51 regions are shown (the number of U.S. states plus Washington DC) but it can easily extended.

Source: (Redrawn from Table 1.2, p. 277 in Symanzik, J. and Carr, D. B. 2008. "Interactive linked micromap plots for the display of geographically referenced statistical data," In *Handbook of Data Visualization*, Chen, C., Hardle, W., and Unwin, A., eds., New York: Springer, pp. 267–294, with kind permission of Springer Science+Business Media.)

References

ACORN Housing/University Partnership. 2007. A peoples' plan for overcoming the Hurricane Katrina blues: A comprehensive strategy for building a more vibrant, sustainable, and equitable 9th ward. Available from http://aap.cornell.edu/aap/crp/outreach/nopi/upload/Peoples_Plan_for_9th_Ward.pdf.

Alberg, A. J., J. G. Ford, and J. M. Samet. 2007. Epidemiology of lung cancer: ACCP evidence-based clinical practice guidelines. *Chest* 132:29–55.

Allison, T., and D. V. Cicchetti. 1976. Sleep in mammals: Ecological and constitutional correlates. *Science* 194:732–34.

Anselin, L., I. Syabri, and Y. Kho. 2006. GeoDa: An introduction to spatial data analysis. *Geographical Analysis* 38:5–22.

Baddeley, A. D. 1981. The cognitive psychology of everyday life. *British Journal of Psychology* 72:257–69.

Bailey, R. G. 1995. *Description of the ecoregions of the United States.* Misc. Publ. 1391, rev. Washington, DC: USDA Forest Service.

Banerjee, S., B. P. Carlin, and A. E. Gelfand. 2004. *Hierarchical modeling and analysis for spatial data.* Boca Raton, FL: Chapman & Hall/CRC.

Bauer, B., P. Jolicoeur, and W. B. Cowan. 1996. Distractor heterogeneity versus linear separability in colour visual search. *Perception* 25:1281–94.

Becker, R. A., and W. S. Cleveland. 1993. Discussion of "Graphical comparisons of several linked aspects" by John W. Tukey. *Journal of Computational and Graphical Statistics* 2:41–48.

Becker, R. A., and W. S. Cleveland. 1996. The design and control of trellis display. *Journal of Computational and Graphical Statistics* 5:123–55.

Becker, R. A., W. S. Cleveland, and G. Weil. 1988. The use of brushing and rotation for data analysis. In *Dynamic graphics for statistics,* ed. W. S. Cleveland and M. E. McGill, 247–75. Pacific Grove, CA: Wadsworth & Brooks/Cole.

Benston, L. 2008. Poker pro urges casino smoking ban. Available from http://www.lasvegassun.com/news/2008/oct/21/poker-pro-urges-casino-smoking-ban/.

Bertin, J. 1973. *Semiologie graphique.* 2. The Hague: Mouton-Gautier.

Bhowmick, T., A. L. Griffin, A. M. MacEachren, B. C. Kluhsman, and E. Lengerich. 2008. Informing geospatial toolset design: Understanding the process of cancer data exploration and analysis. *Health & Place* 14:576–607.

Biello, D. 2007. Searching for God in the brain. *Scientific American Mind,* October.

Brewer, C. A. 2005. *Designing better maps: A guide for GIS users.* Redlands, CA: ESRI Press.

Brewer, C. A. 2006. Basic mapping principles for visualizing cancer data using geographic information systems (GIS). *American Journal of Preventative Medicine* 30:S25–36.

Brewer, C. A., A. M. MacEachren, and L. W. Pickle. 1997. Mapping mortality: Evaluating color schemes for choropleth maps. *Annals of the American Association of Geographers* 87:411–38.

Brewer, C. A., and L. W. Pickle. 2002. Evaluation of methods for classifying epidemiological data on choropleth maps in series. *Annals of the American Association of Geographers* 92:662–81.

Buja, A., D. Cook, and D. F. Swayne. 1996. Interactive high-dimensional data visualization. *Journal of Computational and Graphical Statistics* 5:78–99.

Bureau of Labor Statistics, Census Bureau, and National Cancer Institute. 2009. Current population survey, tobacco use survey. Available from http://riskfactor.cancer.gov/studies/tus-cps/index.html.

Butler, M. A., and C. A. Beale. 1994. *Rural-urban continuum codes for metro and nonmetro counties, 1993.* AGES-9425. Washington, DC: USDA Economic Research Service.

Card, S. K., J. D. Mackinlay, and B. Shneiderman. 1999. *Readings in information visualization: Using vision to think.* San Francisco: Morgan Kaufmann Publishers.

Carr, D. B. 1980. Raster color displays—Examples, ideas and principles. Oakland, CA. In *Proceedings of the 1980 DOE Statistical Symposium,* 116–26.

Carr, D. B. 1991. Looking at large data sets using binned data plots. In *Computing and graphics in statistics,* ed. A. Buja and P. A. Tukey, 7–39. New York: Springer-Verlag.

Carr, D. B. 1994. *Converting tables to plots.* Technical Report 101. Fairfax, VA: Center for Computational Statistics, George Mason University.

Carr, D. B. 2001. Designing linked micromap plots for states with many counties. *Statistics in Medicine* 20:1331–39.

Carr, D. B., J. Chen, B. S. Bell, L. W. Pickle, and Y. Zhang. 2002. Interactive linked micromaps and dynamically conditioned choropleth maps. Los Angeles, CA. In *Proceedings of the Second National Conference on Digital Government Research,* 61–67.

Carr, D. B., R. Kahn, K. Sahr, and A. R. Olsen. 1997. ISEA discrete global grids. *Statistical Computing & Graphics Newsletter* 8:31–39.

Carr, D. B., R. J. Littlefield, W. L. Nicholson, and J. S. Littlefield. 1987. Scatterplot matrix techniques for large N. *Journal of the American Statistical Association* 82:424–36.

Carr, D. B., and W. L. Nicholson. 1988. EXPLOR4: A program for exploring four-dimensional data using stereo-ray glyphs, dimensional constraints, rotation, and masking. In *Dynamic graphics for statistics*, ed. W. S. Cleveland and M. E. McGill, 309–29. Belmont, CA: Wadsworth & Brooks/Cole.

Carr, D. B., W. L. Nicholson, R. J. Littlefield, and D. L. Hall. 1986. Interactive color display methods for multivariate data. In *Statistical image processing and graphics*, ed. E. J. Wegman and D. J. DePriest, 215–50. New York: Marcel Dekker.

Carr, D. B., and A. R. Olsen. 1996. Simplifying visual appearance by sorting: An example using 159 AVHRR classes. *Statistical Computing & Graphics Newsletter* 7:10–16.

Carr, D. B., A. R. Olsen, J. P. Courbois, S. Pierson, and D. A. Carr. 1998a. Linked micromap plots: Named and described. *Statistical Computing & Graphics Newsletter* 9:24–32.

Carr, D. B., A. R. Olsen, S. Pierson, and J. P. Courbois. 1998b. Boxplot variations in a spatial context: An Omernik ecoregion and weather example. *Statistical Computing & Graphics Newsletter* 9(2).

Carr, D. B., A. R. Olsen, S. Pierson, and J. P. Courbois. 2000. Using linked micromap plots to characterize Omernik ecoregions. *Data Mining and Knowledge Discovery* 4:43–67.

Carr, D. B., A. R. Olsen, and D. White. 1992. Hexagon mosaic maps for display of univariate and bivariate geographical data. *Cartography and Geographic Information Systems* 19:228–36, 271.

Carr, D. B., and S. Pierson. 1996. Emphasizing statistical summaries and showing spatial context with micromaps. *Statistical Computing & Graphics Newsletter* 7:16–23.

Carr, D. B., and R. Sun. 1999. Using layering and perceptual grouping in statistical graphics. *Statistical Computing & Graphics Newsletter* 10:25–31.

Carr, D. B., J. F. Wallin, and D. A. Carr. 2000. Two new templates for epidemiology applications: Linked micromap plots and conditioned choropleth maps. *Statistics in Medicine* 19:2521–38.

Carr, D. B., D. White, and A. M. MacEachren. 2005. Conditioned choropleth maps and hypothesis generation. *Annals of the American Association of Geographers* 95:32–53.

Carswell, C. M., H. S. Kinslow, L. W. Pickle, and D. J. Herrmann. 1995. Using color to represent magnitude in statistical maps: The case for double-ended scales. In *Cognitive aspects of statistical mapping*, ed. L. W. Pickle and D. J. Herrmann, 201–28. Hyattsville, MD: National Center for Health Statistics.

Centers for Disease Control and Prevention. 2003. Behavioral Risk Factor Surveillance System. Available from http://www.cdc.gov/brfss.

Centers for Disease Control and Prevention. 2008. Behavioral Risk Factor Surveillance System maps. Available from http://apps.nccd.cdc.gov/gisbrfss/default.aspx.

Centers for Disease Control and Prevention National Center for HIV STD and TB Prevention (NCHSTP) and Division of STD/HIV Prevention. 2005. Sexually transmitted disease morbidity 1984–2003, CDC WONDER On-line database. Available from http://wonder.cdc.gov/std-v2003.html.

Chambers, J. M., W. S. Cleveland, B. Kleiner, and P. A. Tukey. 1983. *Graphical methods for data analysis*. Belmont, CA: Wadsworth International Group.

CIE. 1932. *Commission internationale de l'Eclairage proceedings, 1931*, Cambridge, England.

Cleveland, W. S. 1979. Robust locally weighted regression and smoothing scatterplots. *Journal of the American Statistical Association* 74:829–36.

Cleveland, W. S. 1985. *The elements of graphing data*. Monterey, CA: Wadsworth Advanced Books and Software.

Cleveland, W. S. 1993. *Visualizing data*. Summit, NJ: Hobart Press.

Cleveland, W. S., E. Grosse, and W. M. Shu. 1992. Local regression models. In *Statistical models in S*, ed. J. M. Chambers and T. J. Hastie, 309–72. Pacific Grove, CA: Wadsworth and Brooks/Cole.

Cleveland, W. S., and R. McGill. 1984. Graphical perception: Theory, experimentation, and application to the development of graphical methods. *Journal of the American Statistical Association* 79:531–54.

Cohen, J. 2008. Frustration and optimism in New Orleans. *Washington Post*, August 10, A.3.

Cook, D., and D. F. Swayne. 2007. *Interactive and dynamic graphics for data analysis with R and GGobi*. New York: Springer.

Cubbin, C., L. W. Pickle, and L. Fingerhut. 2000. Social context and geographic patterns of homicide among US black and white males. *American Journal of Public Health* 90:579–87.

Dent, B. D. 1993. *Cartography: Thematic map design*. 3. Dubuque, IA: Wm. C. Brown Publishers.

Devesa, S. S., D. J. Grauman, W. J. Blot, G. A. Pennello, R. N. Hoover, and J. F. Fraumeni, Jr. 1999. *Atlas of cancer mortality in the United States: 1950–94*. NIH Publication 99-4564. Bethesda, MD: National Cancer Institute.

Dickinson, S., H. Christensen, J. Tsotsos, and G. Olofsson. 1997. Active object recognition integrating attention and viewpoint control. *Computer Vision and Image Understanding* 6:239–60.

dos Santos Silva, I. 1999. *Cancer epidemiology: Principles and methods*. 2nd ed. Lyon, France: International Agency for Research on Cancer.

Einstein, A. 2009. Einstein quotes. Available from http://www.heartquotes.net/Einstein.html.

Environmental Systems Research Institute. 2003. *ArcGIS*. Version 8.3. Redlands, CA: ESRI, Inc.

Few, S. 2004. *Show me the numbers*. Oakland, CA: Analytics Press.

Few, S. 2009. *Now you see it*. Oakland, CA: Analytics Press.

Friedman, J. H., and L. C. Rafsky. 1979. Multivariate generalizations of the Walk-Wolfowitz and Smirnov two-sample tests. *Annals of Statistics* 7:697–717.

Friedman, J. H., and J. W. Tukey. 1974. A projection pursuit algorithm for exploratory data analysis. *IEEE Transactions on Computers* C-23:881–90.

Friendly, M. 2006. SAS macro programs: ccmap. Available from http://www.math.yorku.ca/SCS/sasmac/ccmap.html. http://www.math.yorku.ca/SCS/sasmac/ccmap.html.

Friendly, M. 2008. A brief history of data visualization. In *Handbook of data visualization*, ed. C. Chen, W. Hardle, and A. Unwin, 15–56. Berlin: Springer-Verlag.

Friendly, M., and D. J. Denis. 2001. Milestones in the history of thematic cartography, statistical graphics, and data visualization. Available from http://www.math.yorku.ca/SCS/Gallery/milestone/.

Gapminder Foundation. 2009. Gapminder. Available from www.gapminder.org.www.gapminder.org.

Gebreab, S. Y., R. R. Gillies, R. G. Munger, and J. Symanzik. 2008. Visualization and interpretation of birth defects data using linked micromap plots. *Birth Defects Research: Clinical and Molecular Teratology* 82A:110–19.

Gladwell, M. 2000. *The tipping point*. New York: Little, Brown & Company.

Google, Inc. 2009. Google Earth. Available from http://earth.google.com.http://earth.google.com.

Gotway, C. A., and L. J. Young. 2002. Combining incompatible spatial data. *Journal of the American Statistical Association* 97:632–48.

Gould, S. J. 1979. *Ever since Darwin: Reflections in natural history*. New York: W. W. Norton & Company.

Groen, J. A., and A. E. Polivka. 2008. Hurricane Katrina evacuees: Who they are, where they are and how they are faring. *Monthly Labor Review*, March, 32–51. Available from http://www.bls.gov/opub/mlr/2008/03/art3full.pdf.

Hahsler, M., K. Hornik, and C. Buchta. 2008. Getting things in order: An introduction to the R package seriation. *Journal of Statistical Software* 25:1–34.

Hastie, R., O. Hammerle, J. Kerwin, and D. J. Herrmann. 1996. Human performance reading statistical maps. *Journal of Experimental Psychology: Applied* 2:3–16.

Heath, C., and D. Heath. 2007. *Made to stick: Why some ideas survive and others die*. New York: Random House.

Herrmann, D. J., and L. W. Pickle. 1996. A cognitive sub-task model of statistical map reading. *Visual Cognition* 3:165–90.

Herrmann, D. J., C. Y. Yoder, M. Gruneberg, and D. G. Payne. 2006. *Applied cognitive psychology*. Mahwah, NJ: Lawrence Erlbaum Associates.

Heuer, R. J., Jr. 1999. Psychology of intelligence analysis. Available from https://www.cia.gov/library/center-for-the-study-of-intelligence/csi-publications/books-and-monographs/psychology-of-intelligence-analysis/PsychofIntelNew.pdf.

Hill, A. B. 1965. The environment and disease: Association or causation? *Proceedings of the Royal Society of Medicine* 58:295–300.

Ihaka, R., and R. Gentleman. 1996. R: A language for data analysis and graphics. *Journal of Computational and Graphical Statistics* 5:299–314.

Imhof, E. 1982. *Cartographic relief presentation*. Berlin: DeGruyter.

Insightful Corp. *S plus*. 1988. Seattle, WA: Insightful Corp.

Kafadar, K. 1994. Choosing among two-dimensional smoothers in practice. *Computational Statistics & Data Analysis* 18:419–39.

Kahneman, D., A. Treisman, and B. J. Gibbs. 1992. The reviewing of object files: Object-specific integration of information. *Cognitive Psychology* 24:175–219.

Kahneman, D., and A. Tversky. 1979. Prospect theory: An analysis of decision under risk. *Econometrika* 47:263–91.

Keim, D. A. 2002. Information visualization and visual data mining. *IEEE Transactions on Visualization and Computer Graphics* 8:1–8.

Koerber, K. 2006. *Migration patterns and mover characteristics from the 2005 ACS Gulf Coast area special products*. Washington, DC: Census Bureau. Available from http://www.census.gov/Press-Release/www/emergencies/gulf_migration.pdf.

Kosslyn, S. M. 1994a. *Elements of graph design*. New York: W. H. Freeman and Company.

Kosslyn, S. M. 1994b. *Image and brain*. Cambridge, MA: MIT Press.

Kosslyn, S. M. 1995. Mental imagery. In *An Invitation to Cognitive Science*, 2nd ed., Vol. 2, ed. S. M. Kosslyn and D. N. Osherson. 267–296. Cambridge, MA: MIT Press.

Kosslyn, S. M. 2006. *Graph design for the eye and mind*. New York: Oxford University Press.

Kulldorff, M., L. Huang, L. W. Pickle, and L. Duczmal. 2006. An elliptic spatial scan statistic. *Statistics in Medicine* 25:3929–43.

Lewandowsky, S., J. T. Behrens, L. W. Pickle, D. J. Herrmann, and A. A. White. 1995. Perception of clusters in mortality maps: Representing magnitude and statistical reliability. In *Cognitive aspects of statistical mapping*, ed. L. W. Pickle and D. J. Herrmann, 107–32. Hyattsville, MD: National Center for Health Statistics.

Lewandowsky, S., D. J. Herrmann, J. T. Behrens, S.-C. Li, L. W. Pickle, and J. B. Jobe. 1993. Perception of clusters in statistical maps. *Applied Cognitive Psychology* 7:533–51.

Lintott, C. J., K. Schawinski, A. Slosar, K. Land, S. Bamford, D. Thomas, M. J. Raddick, R. C. Nichol, A. Szalay, D. Andreescu, P. Murray, and J. Vandenberg. 2008. Galaxy zoo: Morphologies derived from visual inspection of galaxies from the Sloan Digital Sky Survey. *Monthly Notices of the Royal Astronomical Society* 389:1179–89.

Littlejohn, S. W., and K. A. Koss. 2007. *Theories of human communication*, 9th ed., Belmont, CA: Wadsworth Publishing.

MacEachren, A. M. 1980. Travel time as the basis of cognitive distance. *The Professional Geographer* 32:30–36.

MacEachren, A. M. 1995. *How maps work*. New York: Guilford Press.

MacEachren, A. M., F. P. Boscoe, D. Haug, and L. W. Pickle. 1998. Geographic visualization: Designing manipulable maps for exploring temporally varying georeferenced statistics. Research Triangle Park, NC. In *Proceedings of the IEEE Symposium on Information Visualization 1998*, 87–94.

MacEachren, A. M., C. A. Brewer, and L. W. Pickle. 1998. Visualizing georeferenced data: Representing reliability in health statistics. *Environment and Planning* 30A:1547–61.

MacEachren, A. M., X. Dai, F. Hardisty, D. Guo, and E. Lengerich. 2003a. Exploring high-D spaces with multiform matrices and small multiples. Seattle, WA. In *Proceedings of the IEEE Symposium on Information Visualization 2003*, 31–38.

MacEachren, A. M., F. Hardisty, X. Dai, and L. W. Pickle. 2003b. Supporting visual analysis of federal geospatial statistics. *Communications of the Association of Computing Machinery* 46:63–64.

Maher, R. J. 1995. The interpretation of statistical maps as a function of the map reader's profession. In *Cognitive aspects of statistical mapping*, ed. L. W. Pickle and D. J. Herrmann, 249–74. Hyattsville, MD: National Center for Health Statistics.

Makuc, D. M., B. Haglund, D. D. Ingram, J. C. Kleinman, and J. J. Feldman. 1991. *Health service areas for the United States*. Series 2(112). Hyattsville, MD: National Center for Health Statistics.

Mark, A. 2007. Hochberg and inattentional blindness. In *In the mind's eye: Julian Hochberg on the perception of pictures, films and the world*, ed. M. A. Peterson, B. Gillam, and H. A. Sedgwick, 483–94. New York: Oxford University Press.

Marr, D. 1982. *Vision: A computational investigation into the human representation and processing of visual information*. New York: Henry Holt and Co., Inc.

Mason, T. J. 1995. The development of the series of U.S. cancer atlases: Implications for future epidemiologic research. *Statistics in Medicine* 14:473–79.

Mason, T. J., F. W. McKay, R. N. Hoover, W. J. Blot, and J. F. Fraumeni, Jr. 1975. *Atlas of cancer mortality for U.S. counties: 1950–1969*. Bethesda, MD: U.S. Department of Health, Education, and Welfare.

Mayer, R. E., J. Heiser, and S. Lonn. 2001. Cognitive constraints on multimedia learning: When presenting more material results in less understanding. *Journal of Educational Psychology* 93:187–98.

McCarthy, K., D. J. Peterson, N. Sastry, and M. Pollard. 2006. *The repopulation of New Orleans after Hurricane Katrina*. Santa Monica, CA: The RAND Corporation. Available from www.rand.org/about/katrina.html.

McCloud, S. 1993. *Understanding comics: The invisible art*. New York: Harper Collins Publishers.

McDonald, J. A. 1982. *Interactive graphics for data analysis*. Palo Alto, CA: Stanford Linear Accelerator Center.

Monmonier, M. 1988. Geographical representations in statistical graphics: A conceptual framework. Alexandria, VA. American Statistical Association. In *Proceedings of the Section on Statistical Graphics 1988*, 1–10.

Monmonier, M. 1992. Authoring graphics scripts: Experiences and principles. *Cartography and Geographic Information Systems* 19:247–60.

Monmonier, M. 1993. *Mapping it out: Expository cartography for the humanities and social sciences*. Chicago: University of Chicago Press.

Mungiole, M., L. W. Pickle, and K. H. Simonson. 1999. Application of a weighted head-banging algorithm to mortality data maps. *Statistics in Medicine* 18:3201–9.

National Cancer Institute. 2006. State Cancer Profiles comparative data display [Micromaps]. Availabe from http://statecancerprofiles.cancer.gov/micromaps/.

National Cancer Institute. 2008a. SEER fast stats. Available from http://seer.cancer.gov/faststats.

National Cancer Institute. 2008b. Surveillance, Epidemiology, and End Results (SEER) Program SEER*Stat database: Mortality—All COD, aggregated with state, total U.S. (1969–2005). Available from www.seer.cancer.gov, underlying mortality data provided by NCHS (www.cdc.gov/nchs).

National Cancer Institute Cancer Statistics Branch. 2003. Surveillance, Epidemiology, and End Results (SEER) program populations (1969–2000). Bethesda, MD: National Cancer Institute, DCCPS, Surveillance Research Program. Available from www.seer.cancer.gov/popdata.

Nelson, D. E., B. W. Hesse, and R. T. Croyle. 2009. *Making data talk: Communicating public health data to the public, policy makers, and the press*. New York: Oxford University Press.

Newton, C. 1978. Graphics from alpha to omega in data analysis. In *Graphical representation of multivariate data*, ed. P. C. C. Wang, 59–92. New York: Academic Press.

Oliver, M. N., K. A. Matthews, M. S. Siadaty, F. R. Hauck, and L. W. Pickle. 2005. Geographic bias related to geocoding in epidemiologic studies. *International Journal of Health Geographics* 4:29.

Omernik, J. M. 1987. Ecoregions of the conterminous United States. *Annals of the American Association of Geographers* 77:118–25.

Omernik, J. M. 1995. Ecoregions: A framework for managing ecosystems. *The George Wright Forum* 12:35–51.

Openshaw, S., and P. Taylor. 1979. A million or so correlation coefficients: Three experiments on the modifiable area unit problem. In *Statistical applications in the spatial sciences*, ed. N. Wrigley, 127–44. London: Pion.

Palmer, S. E. 1999. *Vision science: Photons to phenomenology*. Cambridge, MA: MIT Press.

Palmer, S. E., and I. Rock. 1994. Rethinking perceptual organization: The role of uniform connectedness. *Psychonomic Bulletin & Review* 1:29–55.

Pickle, L. W. 2009. A history and critique of U.S. mortality atlases. *Spatial and Spatio-Temporal Epidemiology*. 1:3–17.

Pickle, L. W., and Herrmann, D. J. 1995. *Cognitive aspects of statistical mapping*. Working Paper Series Report 18. Hyattsville, MD: National Center for Health Statistics.

Pickle, L. W., D. J. Herrmann, and B. Wilson. 1995. A legendary study of statistical map reading: The cognitive effectiveness of statistical map legends. In *Cognitive aspects of statistical mapping*, ed. L. W. Pickle and D. J. Herrmann, 233–48. Hyattsville, MD: National Center for Health Statistics.

Pickle, L. W., T. J. Mason, N. Howard, R. N. Hoover, and J. F. Fraumeni, Jr. 1987. *Atlas of U.S. cancer mortality among whites, 1950–1980*. Washington, DC: U.S. Government Printing Office.

Pickle, L. W., M. Mungiole, G. K. Jones, and A. A. White. 1996. *Atlas of United States mortality*. Hyattsville, MD: National Center for Health Statistics.

Pickle, L. W., and A. A. White. 1995. Effects of the choice of age-adjustment method on maps of death rates. *Statistics in Medicine* 14:615–27.

Pink, D. H. 2005. *A whole new mind: Why right-brainers will rule the future*. New York: Riverhead Books.

Pirolli, P. 2003. A theory of information scent. In *Human-computer interaction*, ed. J. Jacko and C. Stephanidis, 213–17. Mahwah, NJ: Lawrence Erlbaum.

Playfair, W. 1786. *The commercial and political atlas*. 1. London: J. Wallis.

Pylyshyn, Z. W. 1973. What the mind's eye tells the minds brain. *Psychological Bulletin* 80:1–24.

R Development Core Team. 2009. *R: A language and environment for statistical computing*. Vienna, Austria: R Foundation for Statistical Computing. Available from http://www.R-project.org.

Rensink, R. A. 2000. The dynamic representation of scenes. *Visual Cognition* 7:17–42.

Rensink, R. A. 2002. Change detection. *Annual Review of Psychology* 53:245–77.

Rensink, R. A., J. K. O'Regan, and J. J. Clark. 1997. To see or not to see: The need for attention to perceive changes in scenes. *Psychological Science* 8:368–73.

Reynolds, G. 2008. *Presentation zen: Simple ideas on presentation design and delivery*. 1st ed. Gresham, OR: New Riders Publishing.

Roam, D. 2008. *The back of the napkin*. New York: Portfolio.

Robertson, G., R. Fernandez, D. Fisher, B. Lee, and J. T. Stasko. 2008. Effectiveness of animation in trend visualization. *IEEE Transactions on Visualization and Computer Graphics* 14:1325–32.

Robinson, A. C., J. Chen, E. Lengerich, H. Meyer, and A. M. MacEachren. 2005. Combining usability techniques to design geovisualization tools for epidemiology. *Cartography and Geographic Information Science* 32:243–55.

Robinson, W. S. 1950. Ecological correlations and the behavior of individuals. *American Sociological Review* 15:351–57.

Sagan, C. 1977. *The dragons of Eden: Speculations on the evolution of human intelligence*. New York: Random House.

SAP. 2008. *TableLens*. Ver. 3.1. Available from http://www.sap.com/solutions/sapbusinessobjects/large/intelligenceplatform/bi/dashboard-visualization/advanced-visualization/index.epx.

Schabenberger, O., and C. A. Gotway. 2005. *Statistical methods for spatial data analysis*. Boca Raton, FL: Chapman & Hall/CRC.

Scott, D. W. 1992. *Multivariate density estimation, theory, practice and visualization*. New York: John Wiley & Sons.

Shneiderman, B. 1992. *Designing the user interface: Strategies for effective human-computer interaction*. 2nd ed. Reading, MA: Addison Wesley.

Simon, H. A. 1996. *The sciences of the artificial*. 3rd ed. Cambridge, MA: MIT Press.

Simons, D. J., and C. F. Chabris. 1999. Gorillas in our midst: Sustained inattentional blindness for dynamic events. *Perception* 28:1059–74.

Stafford, T., and M. Webb. 2004. *Mind hacks: Tips and tricks for using your brain*. Sebastopol, CA: O'Reilly.

Statistical Research and Applications Branch and National Cancer Institute. 2009. Linked micromaps program. Version 1.02. Available from http://gis.cancer.gov/tools/micromaps/.

Sternberg, S. 1975. Memory scanning: New findings and current controversies. *Quarterly Journal of Experimental Psychology* 27:1–32.

Swayne, D. F., D. Cook, and A. Buja. 1998. XGobi: Interactive dynamic graphics in the X Window System. *Journal of Computational and Graphical Statistics* 7:113–30.

Swayne, D. F., D. T. Lang, A. Buja, and D. Cook. 2003. GGobi: Evolving from XGobi into an extensible framework for interactive data visualization. *Computational Statistics & Data Analysis* 43:423–44.

Symanzik, J., D. A. Axelrad, D. B. Carr, J. Wang, D. Wong, and T. J. Woodruff. 1999. HAPs, micromaps and GPL—Visualization of geographically referenced statistical summaries on the World Wide Web. The Annual Proceedings of the American Congress on Surveying and Mapping.

Symanzik, J., and D. B. Carr. 2008. Interactive linked micromap plots for the display of geographically referenced statistical data. In *Handbook of data visualization*, ed. C. Chen, W. Hardle, and A. Unwin, 267–94. New York: Springer.

Symanzik, J., D. Cook, N. Lewin-Koh, J. J. Majure, and I. Megretskaia. 2000. Linking ArcView and XGobi: Insight behind the front end. *Journal of Computational and Graphical Statistics* 9:470–90.

Takatsuka, M., and M. Gahegan. 2002. GeoVISTA Studio: A codeless visual programming environment for geoscientific data analysis and visualization. *Computers and Geosciences* 28:1131–44.

Thomas, J. J. 2007. Visual analytics: Why now? *Information Visualization* 6:104–6.

Thomas, J. J., and K. A. Cook. 2005. *Illuminating the path: The research and development agenda for visual analytics.* Washington, DC: Department of Homeland Security, National Visualization and Analytics Center.

TIBCO Software, Inc. 2008. Spotfire enterprise analytics platform. Available from http://spotfire.tibco.com/index.cfm.

Tobler, W. 1970. A computer movie simulating urban growth in the Detroit region. *Economic Geography* 46:234–40.

Tufte, E. R. 1983. *The visual display of quantitative information.* Cheshire, CT: Graphics Press.

Tufte, E. R. 1990. *Envisioning information.* Cheshire, CT: Graphics Press.

Tukey, J. W. 1977. *Exploratory data analysis.* Reading, MA: Addison-Wesley Publishing Company.

Tukey, J. W. 1979. Statistical mapping: What should not be plotted. In *Proceedings of the 1976 Workshop on Automated Cartography,* Hyattsville, MD, March 18, 18–26.

Tukey, J. W. 1988a. Comment. In *Dynamic graphics for statistics,* ed. W. S. Cleveland and M. E. McGill, 50–54. Pacific Grove, CA: Wadsworth and Brooks/Cole.

Tukey, J. W. 1988b. Foreword. In *The collected works of John W. Tukey: Graphics, 1965–1985,* ed. W. S. Cleveland, xxxv. Pacific Grove, CA: Wadsworth Advanced Books and Software.

Tukey, J. W., and P. A. Tukey. 1985. Computer graphics and exploratory data analysis: An introduction. Dallas, TX. In *Proceedings of the Sixth Annual Conference and Exposition: Computer Graphics '85,* Vol. III, 773–85.

Tukey, P. A., and J. W. Tukey. 1981. Graphic display of data sets in 3 or more dimensions. In *Interpreting multivariate data,* ed. V. Barnett, 189–275. Chichester, UK: John Wiley.

Tversky, A., and D. Kahneman. 1974. Judgment under uncertainty: Heuristics and biases. *Science* 185:1124–31.

Tversky, B. 1981. Distortions in memory for maps. *Cognitive Psychology* 13:407–33.

Tversky, B. 2005. Functional significance of visuospatial representations. In *The Cambridge handbook of visuospatial thinking,* ed. P. Shah and A. Miyake, 1–34. New York: Cambridge University Press.

Tversky, B., and D. Schiano. 1989. Perceptual and conceptual factors in distortions in memory for maps and graphs. *Journal of Experimental Psychology: General* 118:387–98.

U.S. Bureau of the Census. 1994. *Geographic areas reference manual.* Available from www.census.gov/geo/www/garm.html.

U.S. Department of Health and Human Services. 2000. *Healthy people 2010. With understanding and improving health and objectives for improving health.* 2nd ed. Washington, DC. Available from www.health.gov/healthypeople.

Van den Berg, B. J. 1977. Epidemiologic observations of prematurity: Effects of tobacco, coffee and alcohol. In *The epidemiology of prematurity,* ed. D. M. Reed and F. J. Stanley, 157–76. Baltimore: Urban and Schwarzenberg.

Wainer, H. 1993. Tabular presentation. *Chance* 6:52–56.

Wainer, H., and C. Francolini. 1980. An empirical inquiry concerning human understanding of two-variable color maps. *The American Statistician* 34:81–93.

Waller, L. A., and C. A. Gotway. 2004. *Applied spatial statistics for public health data.* New York: John Wiley & Sons.

Ware, C. 2004. *Information visualization: Perception for design.* 2nd ed. San Francisco: Morgan Kaufmann Publishers.

Ware, C. 2008. *Visual thinking for design.* Burlington, MA: Morgan Kaufmann Publishers.

Ware, C., and P. Mitchell. 2008. Visualizing graphs in three dimensions. *ACM Transactions on Applied Perception* 5:1–15.

Wason, P. C. 1960. On the failure to eliminate hypotheses in a conceptual task. *Quarterly Journal of Experimental Psychology* 12:129–40.

Weaver, C. 2004. Building highly-coordinated visualizations in improvise. Austin, TX. In *Proceedings of the IEEE Symposium on Information Visualization 2004,* 159–66.

Weaver, C., D. Fyfe, A. H. Robinson, D. Holdsworth, D. Peuquet, and A. M. MacEachren. 2007. Visual exploration and analysis of historic hotel visits. *Information Visualization* 6:89–103.

Wegman, E. J. 1990. Hyperdimensional data analysis using parallel coordinates. *Journal of the American Statistical Association* 85:664–75.

Weisberg, S. 1980. *Applied linear regression.* New York: John Wiley & Sons.

White, D., E. M. Preston, K. E. Freemark, and A. R. Kiester. 1999. A hierarchical framework for conserving biodiversity. In *Landscape ecological analysis: Issues and applications,* ed. J. M. Klopatek and R. H. Gardner, 127–53. New York: Springer-Verlag.

White, D., and J. C. Sifneos. 2002. Regression tree cartography. *Journal of Computational and Graphical Statistics* 11:600–14.

Whittaker, G., and D. W. Scott. 1994. Spatial estimation and presentation of regression surfaces in several variables via the averaged shifted histogram. *Computing Science and Statistics* 26:8–17.

Wikipedia. 2009. 2005 Atlantic hurricane season. Available from http://en.wikipedia.org/wiki/2005_Atlantic_hurricane_season.

Wilk, M. B., and R. Gnanadesikan. 1968. Probability plotting methods for the analysis of data. *Biometrika* 55:1–17.

Wilkinson, L. 1984. *SYSTAT: The system for statistics*. Evanston, IL: SYSTAT, Inc.

Wilkinson, L. 1999. *The grammar of graphics*. 1st ed. New York: Springer.

Wilkinson, L. 2005. *The grammar of graphics*. 2nd ed. New York: Springer.

Wilkinson, L., A. Anand, and R. Grossman. 2005. Graph-theoretic scagnostics. *IEEE Symposium on Information Visualization* 2005:157–64.

Wilkinson, L., and G. Wills. 2008. Scagnostics distributions. *Journal of Computational and Graphical Statistics* 17:473–91.

Winn, D. M., W. J. Blot, C. M. Shy, L. W. Pickle, A. Toledo, and J. F. Fraumeni, Jr. 1981. Snuff dipping and oral cancer among women in the southern United States. *New England Journal of Medicine* 304:745–49.

Index